Lecture Notes in Mathematics

Edited by A. Dold and B. Eckmann

879

Vjačeslav V. Sazonov

Normal Approximation – Some Recent Advances

Springer-Verlag
Berlin Heidelberg New York 1981

Author

Vjačeslav V. Sazonov
Steklov Institute of Mathematics
42 Vavilov Street, 117966 Moscow, USSR

AMS Subject Classifications (1980): 60B12, 60E10, 60E99, 60F05

ISBN 3-540-10863-7 Springer-Verlag Berlin Heidelberg New York
ISBN 0-387-10863-7 Springer-Verlag New York Heidelberg Berlin

Printed in Germany

Printing and binding: Beltz Offsetdruck, Hemsbach/Bergstr.
2141/3140-543210

Preface

This course is devoted to the problem of estimation of the speed of convergence in the central limit theorem in R^k and in Hilbert space H. If $X_1, \ldots, X_n$ are independent identically distributed random variables with values in R^k or H and with $E|X_1|^2 < \infty$, P_n is the distribution of $n^{-1/2}\sum_1^n (X_j - EX_j)$ and Q is the normal probability measure with the same first and second moments as X_1 then the central limit theorem states that for a large class of functions f

$$\int f(x)(P_n - Q)(dx) \qquad (*)$$

tends to zero as $n \to \infty$; the problem is to construct upper bounds for the absolute value of $(*)$. This topic in the finite dimensional case was also covered in the recent monograph [8]. The difference between our approach and that of [8] is that we use mainly a method depending directly on convolutions rather then the method of characteristic functions. For the first time the method of convolutions in its explicit form was used by H.Bergström in 1944. In a number of respects it is simplier then the method of characteristic functions and leads to the goal more directly. We illustrate the method of convolutions in a simple example in §1 of Chapter I. In deriving the integral type estimates however we also apply the method of characteristic functions since we do not have proofs based on the method of convolutions only. Most of our estimates are expressed not in terms of absolute moments as in [8] but in terms of pseudomoments (for definitions see §1,2 of Chapter I). This makes the estimates sharper, the pseudomoments being distances from the distribution of summands to the corresponding normal measure. When the pseudomoments are small enough - less then some absolute constant - estimates for the variation distance are obtained which have the usual order $n^{-1/2}$. Special attention is paid to the constants involved in the estimates: in the k-dimensional case upper bounds of the form ck^m are obtained, where c is an absolute constant and m is a constant corresponding to the type of the estimate.

We do not try to prove results in full generality or to cover all the known achievements in the area. Our main aim is to present the main directions and methods (with the emphasis on the method of convolutions).

The Notes were written at the time the author was lecturing at UCLA in the spring of 1979 and at Moscow State University in 1979-1980. In the lectures Chapter I and §1 of Chapter II were covered (at UCLA Chapter I only). The content of Chapter II, §2 was also planned but was not delivered owing to lack of time.

I am greatly indebted to Professor Yu.V.Prohorov for his valuable advice and criticism during my work in this field. I am also grateful to Professor A.V.Balakrishnan for inviting me to UCLA and for his suggestion to write these Notes. While I was at UCLA we had many inspiring conversations about the theory of probability.

My special thanks go to Miss Stella Lozano of UCLA for typing a part of the first draft and to Mr. V.P.Sharma of ISI, Delhi and to Wendy Coutts of UBC, Vancouver for typing the final version of the manuscript.

V.V.Sazonov

Table of Contents

The Main Notations

R^k	standard k-dimensional Euclidean space
H	a real separable Hilbert space
(x,y)	for $x,y \in R^k$ or H is the inner product of x and y $\left((x,y) = \sum_1^k x_j y_j,\ \text{for } x = (x_1,\dots,x_k),\ y = (y_1,\dots,y_k) \in R^k\right)$
$\lvert x\rvert = (x,x)^{\frac{1}{2}}$	
$\lVert a\rVert = \sum_1^k a_j$	for a nonnegative integral vector $a = (a_1,\dots,a_k) \in R^k$ (i.e. with nonnegative integer components a_j)
$a!$	for a nonnegative integral vector $a \in R^k$ is $a_1!a_2! \dots a_k!$.
D_a	for a nonnegative integral vector $a \in R^k$ is the differential operator $\partial^{\lVert a\rVert}/\partial x_1^{a_1} \dots \partial x_k^{a_k}$
x^a	for $x \in R^k$ and a nonnegative integral vector $a \in R^k$ is $x_1^{a_1} \dots x_k^{a_k}$
$x \le t$	for $x = (x_1,\dots,x_k),\ y = (y_1,\dots,y_k) \in R^k$ means $x_j \le y_j,\ j = \overline{1,k}$
$x < y$	for $x,y \in R^k$ means $x \le y,\ x \ne y$
$\lvert V\rvert$	for a $k \times k$ matrix V is its determinant
$\lVert V\rVert$	for a $k \times k$ matrix is its norm, i.e. $\lVert V\rVert = \sup_{\lvert x\rvert \le 1} \lvert Vx\rvert$
I	identity $k \times k$ matrix
$S_r(x)$	open ball of radius r with centre at x
$S_{r,\varepsilon}(x)$	$= S_{r+\varepsilon}(x) \setminus S_r(x)$
χ_E	indicator function of a set E, i.e. $\chi_E(x) = 1$ or 0 according to $x \in E$ or $x \notin E$
E^c	complement of a set E
$\bar{E}$	closure of a set E
∂E	boundary of a set E
$E^{\varepsilon} = \cup\{S_{\varepsilon}(x) : x \in E\}$	
$E^{-\varepsilon} = \cup\{x : S_{\varepsilon}(x) \subset E\}$	
C	class of all convex Borel subsets of R^k
P_X	probability measure corresponding to a random variable X
$\hat{K}$	for a probability measure or a distribution function K is its characteristic function, $\hat{K}(x) = \int \exp\{i(x,y)\}K(dy)$

$\|M\|$	for a signed measure M denotes its variation, i.e. $\|M\| = M^+ + M^-$, where M^+, M^- are the components of Jordan-Hahn decomposition of M
K^n	where K is a distribution function or a probability measure is the n-fold convolution of K with itself
$N_{\mu,V}$	normal distribution in R^k with mean μ and covariance matrix V
$N_T, T > 0$	is $N_{0,T^{-2}I}$
N =	N_1
Φ	is normal $(0,1)$ distribution function
$\phi_{\mu,V}$ (resp. ϕ_T, ϕ)	density function corresponding to $N_{\mu,V}$ (resp. N_T, N)
c (resp. $c(.)$)	with or without indices denote positive constants (resp. positive constants depending only on quantities in parentheses); the same symbol may stand for different constants.

If the area of an integration or the set over which $\max_x$, $\inf_x$, etc. is taken is not indicated it is understood that the integration is over the whole space and $\max_x$, $\inf_x$, etc. is with respect to the all possible values of x.

CHAPTER I

THE FINITE DIMENSIONAL CASE

Let $X_1, X_2, \dots$ be i.i.d. random variables with values in R^k such that $E|X_1|^2 < \infty$ and suppose that the covariance matrix V of X is nondegenerate. Then by the central limit theorem the distributions P_n of the normalized sums $n^{-\frac{1}{2}} \sum_1^n V^{-\frac{1}{2}}(X_i-\mu)$, where $\mu = EX_1$, converge weakly to the standard normal distribution N on R^k, i.e.

$$\int f(x)P_n(dx) \to \int f(x)N(dx) \quad \text{as} \quad n \to \infty$$

for any bounded Borel measurable function f on R^k such that the set of points of its discontinuity has N-measure zero. We are interested in the estimation of the speed of this convergence. In this Chapter we shall construct bounds for $\int f(x)(P_n-N)(dx)$ for any bounded Borel measurable function on R^k, which depend in general on n, the function f and on moments or pseudomoments of X_1 up to some order. Of special interest is the particular case when f is the indicator of a set E, i.e. $f(x) = \chi_E(x)$. The weak convergence $P_n \Rightarrow N$ itself imply the uniformity of the convergence $(P_n-N)(E) \to 0$ over large classes $\mathcal{E}$ of sets E, e.g. over the class $\mathcal{C}$ of all convex Borel sets, and some of our bounds for $|(P_n-N)(E)|$ for convex sets will be independent of particular sets.

§ 1. The two main methods and the main directions of extensions and improvements of classical results.

We start with an illustration of the two main methods employed in the estimation of the speed of convergence in the central limit theorem in R^k, i.e. the method of compositions and the method of characteristic functions. To this aim we shall give two different proofs of the classical Berry-Esseen theorem (one dimensional case) using these two different methods.

The Berry-Esseen theorem Let $X_1, X_2, \dots$ be independent identically distributed (i.i.d.) real random variables with $EX_1 = \mu, E(X_1-\mu)^2 = \sigma^2 < \infty$. Denote

$$F_n(x) = P(\sigma^{-1} n^{-\frac{1}{2}} \sum_1^n (X_r-\mu) < x)$$

and let Φ be the standard normal distribution function on R. Then for all n = 1,2,...

$$\sup_x |F_n(x) - \Phi(x)| \leq \bar{c}\, \beta_3 \sigma^{-3}\, n^{-\frac{1}{2}} \tag{1}$$

where $\beta_3 = E|X_1-\mu|^3$ and $\bar{c}$ is an absolute constant.

Both proofs we are going to give for this theorem use the following smoothness lemma.

Lemma 1. Let F, G, H be three distribution functions. Assume that

$$H(v) - H(-v+0) \geq u \quad \text{for some} \quad v > 0,\ 2^{-1} < u \leq 1, \tag{2}$$

$$|G(x_1) - G(x_2)| \leq M|x_1 - x_2| \quad \text{for some} \quad M > 0 \ \text{and all} \ x_1, x_2 \in R,$$

and denote

$$D(x) = F(x) - G(x), \qquad H_T(x) = H(Tx),\ T > 0$$

$$\delta = \sup_x |D(x)| \quad , \qquad \delta_T = \sup_x |D*H_T(x)|.$$

Then for any $T > 0$

$$\delta \leq (2u-1)^{-1}\, [\delta_T + 2uv\, MT^{-1}] \tag{3}$$

Proof. Take a $T > 0$ and put $t = vT^{-1}$. Clearly either $\delta = -\inf D(x)$ or $\delta = \sup_x D(x)$. We assume the former case. The latter case is treated similarly. For any $\varepsilon > 0$ there exist x_ε such that $D(x_\varepsilon) < -\delta+\varepsilon$. Note that if $|y| < t$, then

$$\begin{aligned} D(x_\varepsilon - t - y) &\leq (F(x_\varepsilon) - G(x_\varepsilon)) - (G(x_\varepsilon - t-y) - G(x_\varepsilon)) \\ &\leq -\delta + \varepsilon + 2Mt \ . \end{aligned}$$

Therefore

$$\begin{aligned} -\delta_T \leq D*H_T(x_\varepsilon - t) &= \Big(\int_{|y|<t} + \int_{|y|\geq t} \Big) D(x_\varepsilon - t-y)\,dH_T(y) \\ &\leq (-\delta+\varepsilon+2Mt)\lambda + \delta(1-\lambda), \end{aligned}$$

where $\lambda = H_T(t) - H_T(-t+0) = H(v) - H(-v+0)$. Since ε may be taken arbitrary small and $\lambda \geq u$ (see (2)), we have

$$\delta \leq (2\lambda-1)^{-1}(\delta_T + 2tM\lambda) \leq (2u-1)^{-1}(\delta_T + 2uvMT^{-1}). \ \square$$

Corollary 1. Let in lemma 1 $G = H = \Phi$, where Φ is the standard normal distribution. In this case $M = (2\pi)^{-\frac{1}{2}}$ and we may take e.g. $u = 3/4$. Then $v = 1.16$ and we have for these G and H

$$\delta \leq 2\delta_T + (7/2\sqrt{2\pi})T^{-1}. \tag{4}$$

<u>Corollary 2</u>. Let F,G be distribution functions. Suppose G satisfies (2) for some $M > 0$. Then for any $u: 2^{-1} < u < 1$ there exist a positive number $v(u)$ such that for all $T > 0$

$$\sup_x |F(x)-G(x)| \leq (2u-1)^{-1}[(2\pi)^{-1} \int_{-T}^{T} |(\hat{F}(t)-\hat{G}(t))/t|dt+2uv(u)MT^{-1}] \tag{5}$$

<u>Proof</u>. Let H(x) be the distribution function with density $p(x) = (1-\cos x)/\pi x^2$; the corresponding characteristic function is

$$\hat{H}(t) = \begin{cases} 1-|t| , & |t| \leq 1 \\ 0 , & |t| > 1 . \end{cases}$$

The characteristic function corresponding to $H_T(x)$ is $\hat{H}(tT^{-1})$; it vanishes outside the interval $[-T,T]$.

By the inversion formula we have for any distribution function S

$$S*H_T(x)-S*H_T(y) = (2\pi)^{-1} \int_{-T}^{T} (e^{-ity}-e^{-itx})(it)^{-1}\hat{S}(t)\hat{H}_T(t)dt. \tag{6}$$

Suppose that

$$\int_{-T}^{T} |(\hat{F}(t)-\hat{G}(t))/t|dt < \infty . \tag{7}$$

Then by the Riemann-Lebesgue theorem,

$$\int_{-T}^{T} (\hat{F}(t)-\hat{G}(t))t^{-1}\hat{H}_T(t)e^{-ity}dt \to 0 \quad \text{as} \quad y \to \infty . ,$$

Putting in (6) at first S=F and then S=G, substracting the obtained inequalities and letting $y \to \infty$ we thus have

$$F*H_T(x)-G*H_T(x) = (2\pi)^{-1} \int_{-T}^{T} e^{-itx}(\hat{G}(t)-\hat{F}(t))(it)^{-1}\hat{H}_T(t)dt . \tag{8}$$

From (8) it follows immediately that

$$|(F-G)*H_T(x)| \leq (2\pi)^{-1}\int_{-T}^{T}|(\hat{F}(t)-\hat{G}(t))/t|dt . \tag{9}$$

Obviously (9) is true even if inequality (7) fails. Define now $v(u)$, $2^{-1} < u < 1$, by the equation $2H(-v(u)) = u$. The inequality (5) is now an immediate corollary of lemma 1 and (9). □

<u>Proof 1 of the Berry-Esseen theorem</u> (The method of compositions): It is enough to deal with the case $\mu = 0, \sigma = 1$) (To reduce the general case to this one it is enough to consider $Y_j = (X_j-\mu)/\sigma$).

Noting that $\beta_3 \geq (EX_1^2)^{3/2} = 1$, we have

$$|F_1(x)-\Phi(x)| \leq \beta_3 \quad ,$$

thus (1) is true for $n = 1$. We shall show that if (1) is true for all values of n less then some fixed value, then it is also true for this fixed value of n with the same constant $\bar{c}$ if $\bar{c}$ is large enough. In what follows we shall assume $n \geq 2$.

Denote by $F_{(n)}$ the distribution function of $X_1 n^{-\frac{1}{2}}$ and let $\Phi_T(x) = \Phi(Tx), T > 0$. Then $F_n = F^n_{(n)}$ and $\Phi = \Phi_1 = \Phi^n_{n^{-\frac{1}{2}}}$. The following equation is the very essential point of the method of compositions

$$
\begin{aligned}
(F_n-\Phi)*\Phi_T &= (F^n_{(n)} - \Phi^n_{n^{-\frac{1}{2}}})*\Phi_T \\
&= \sum_{i=0}^{n-1} (F^i_{(n)} * \Phi^{n-i-1}_{n^{-\frac{1}{2}}})*\Phi_T*(F_{(n)}-\Phi_{n^{-\frac{1}{2}}}) \\
&=[\sum_{i=1}^{n-1} (F^i_{(n)}-\Phi^i_{n^{-\frac{1}{2}}})*\Phi^{n-i-1}_{n^{-\frac{1}{2}}}*\Phi_T+n\Phi^{n-1}_{n^{-\frac{1}{2}}} *\Phi_T]*(F_{(n)}- \Phi_{n^{-\frac{1}{2}}}) \\
&= (\sum_{i=1}^{n-1} U_i + n \, \Phi_{\tau_0})*H_1 \quad ,
\end{aligned} \tag{10}
$$

where

$$U_i=H_i*\Phi_{\tau_i} \; , \; H_i = F^i_{(n)} - \Phi^i_{n^{-\frac{1}{2}}} \; , \; \tau_i = ((n-i-1)/n+T^{-2})^{-\frac{1}{2}}$$

Representing $U_i(x-y)$ in the form

$$
\begin{aligned}
U_i(x-y) &= \int H_i(x-y-u)\phi_{\tau_i}(u)du \\
&= \int H_i(x-z)\phi_{\tau_i}(y-z)dz
\end{aligned}
$$

where ϕ_T is the density function of Φ_T, we see that it is a smooth function of y and we can write

$$U_i(x-y) = \sum_{j=0}^{2} (-1)^j(j!)^{-1}U_i^{(j)}(x)y^j-6^{-1}U_i'''(x+\theta y)y^3, \tag{11}$$

where $|\theta| \leq 1$. Note that

$$\int y^j \, dH_1(y) = 0, \; j = 0,1,2,$$

since $F_{(n)}$ and $\Phi_{n^{-\frac{1}{2}}}$ both have mean value 0 and variance $1/n$. Therefore

$$
\begin{aligned}
|U_i*H_1(x)| &= | \int U_i(x-y)dH_1(y)| \\
&= 6^{-1}|\int U_i'''(x+\theta y)y^3 dH_1(y)| \\
&\leq 6^{-1} \sup_x |U_i'''(x)|\int|y|^3(dF_{(n)}(y) + d\Phi_{n^{-\frac{1}{2}}}(y)) \quad .
\end{aligned}
$$

The definitions of $F_{(n)}$ and Φ_T imply that

$$\int |y|^3 dF_{(n)}(y) = \beta_3 n^{-3/2}, \int |y|^3 d\Phi_{n^{\frac{1}{2}}}(y) = (4/2\pi) n^{-3/2} ,$$

hence

$$\int |y|^3 (dF_{(n)}(y) + d\Phi_{n^{\frac{1}{2}}}(y)) \leq c\beta_3 n^{-3/2}$$

(recall that $\beta_3 \geq 1$). On the other hand

$$\sup_x |U_i'''(x)| = \sup_x |\int H_i(-z)\phi_{\tau_i}'''(x+z)dz|$$

$$\leq \sup_z |H_i(z)| \int |\phi_{\tau_i}'''(z)| dz .$$

A simple calculation shows that

$$\int |\phi_{\tau_i}'''(z)| dz \leq c\tau_i^3 ,$$

and by the inductive hypothesis we have, since $\Phi^i_{n^{\frac{1}{2}}}$ is normal $(0, i/n)$,

$$\begin{aligned}|H_i(z)| &= |P(n^{-\frac{1}{2}} \sum_1^i X_j < z) - \Phi^i_{n^{\frac{1}{2}}}(z)| \\ &= |F_i((n/i)^{\frac{1}{2}} z) - \Phi((n/i)^{\frac{1}{2}} z)| \\ &\leq \bar{c}\beta_3\, i^{-\frac{1}{2}}. \end{aligned} \tag{12}$$

Using the above inequalities and noting that

$$\begin{aligned}\sum_{i=1}^{n-2} \tau_i^3\, i^{-\frac{1}{2}} &\leq \int_1^{n-1} \frac{dx}{(x-1)^{\frac{1}{2}}((n-x-1)/n + T^{-2})^{3/2}} \\ &= 2n^{\frac{1}{2}} T \frac{((n-2)/n)^{\frac{1}{2}}}{((n-2)/n) + T^{-2}} \\ &\leq 2\sqrt{3}\, n^{\frac{1}{2}}\, T, \end{aligned} \tag{13}$$

we obtain

$$\sum_{i=1}^{n-2} |U_i * H_1(x)| \leq \bar{c}\, c\beta_3^2\, Tn^{-1} \tag{14}$$

For $i = n-1$ by the same argument but using representation

$$U_{n-1}(x-y) = U_{n-1}(x) - U'_{n-1}(x+\theta y)y$$

instead of (11) we deduce

$$|U_{n-1} * H_1(x)| \leq \bar{c}\, c\beta_3^2 \beta T n^{-1} \tag{15}$$

For the term $|\Phi_{\tau_0} * H_1(x)|$ we have (using again the above arguments)

$$|\Phi_{\tau_0} * H_1(x)| = | \int\Phi_{\tau_0}(x-y)dH_1(y)|$$

$$= \frac{1}{6} |\int\Phi'''_{\tau_0}(x+\theta y)y^3 \, dH_1(y)|$$

$$\leq \frac{1}{6} \sup_y |\Phi'''_{\tau_0}(y)| \int|y|^3 \, d(F_{(n)}(y)+\Phi_{n^{\frac{1}{2}}}(y))$$

$$\leq c\beta_3 \, n^{-3/2} \tag{16}$$

since

$$\sup_y |\Phi'''_{\tau_0}(y)| = \tau_0^3 \sup_y|\Phi'''(\tau_0 y)|$$

and

$$\tau_0 = ((n-1)n^{-1} + T^{-2})^{-\frac{1}{2}} \leq \sqrt{2}$$

when $n \geq 2$. Hence we have, combining (10), (14) - (16),

$$|(F_n-\Phi)*\Phi_T(x)| \leq c(\bar{c}\beta_3^2 \, Tn^{-1} + \beta_3 \, n^{-\frac{1}{2}}) ,$$

which implies, by Corollary 1, for any $T > 0$

$$|F_n(x)-\Phi(x)| \leq c(\bar{c}\beta_3^2 \, Tn^{-1} + \beta_3 \, n^{-\frac{1}{2}} + T^{-1}). \tag{17}$$

The right side of (17) attains its minimum when $T = \beta_3^{-1}(n/\bar{c})^{\frac{1}{2}}$ and with this value of T (17) takes the form

$$|F_n(x)-\phi(x)| \leq \bar{c}\beta_3 \, n^{-\frac{1}{2}} \, c(\bar{c}^{-1} + 2\bar{c}^{-\frac{1}{2}}) .$$

Assuming now that $\bar{c}$ is chosen large enough to satisfy $c(\bar{c}^{-1} + 2\bar{c}^{-\frac{1}{2}}) \leq 1$ we obtain (1).□

For the second proof of the Berry-Esseen theorem we will also need the following lemma.

Lemma 2. Let X be a random variable with distribution function F and suppose that $EX = 0$, $EX^2 = 1, \beta_3 = E|X|^3 < \infty$. If

$$|t| < n^{\frac{1}{2}}/5\beta_3 \tag{18}$$

then

$$|\hat{F}^n(tn^{-\frac{1}{2}}) - e^{-t^2/2}| \leq (7/6)\int|t|^3 n^{-\frac{1}{2}} \, \beta_3 e^{-t^2/4}$$

Proof. Since $\beta_3 < \infty$, $\hat{F}$ is three times continuously differentiable and $\hat{F}^{(j)}(u) = \int e^{iux} x^j \, dF(x), j = \overline{1,3}$. Moreover we can write

$$\hat{F}(u) = \sum_{j=0}^{2} \frac{u^j}{j!} \hat{F}^{(j)}(0) + \frac{u^3}{2} \int_0^1 (1-v)^2 \, \hat{F}'''(vu)dv \tag{19}$$

which implies

$$\hat{F}(tn^{-\frac{1}{2}}) = 1-t^2/(2n) + \theta 6^{-1}t^3\beta_3\, n^{-3/2}$$

where $|\theta| \leq 1$. If t satisfies (17) then

$$t^2/(2n) + 6^{-1}|t|^3\beta_3 n^{-3/2} \leq 25^{-1}(\text{since } \beta_3 \geq \sigma^3 = 1),$$

and, consequently, $|f(tn^{-\frac{1}{2}})| \geq 24/25 > 0$. For such t we can thus write

$$\hat{F}^n(tn^{-\frac{1}{2}}) = \exp\{n \log \hat{F}(tn^{-\frac{1}{2}})\}.$$

Furthermore, expanding $\log \hat{F}$ in the same way as $\hat{F}$ in (19) we have

$$\log \hat{F}(tn^{-\frac{1}{2}}) = - t^2/(2n) + \theta 6^{-1}t^3 n^{-3/2}(\log \hat{F})'''(\theta_1 tn^{-\frac{1}{2}})$$

where

$$\begin{aligned}|(\log \hat{F})'''| &= |(\hat{F}'''\, \hat{F}^2 - 3\hat{F}''\, \hat{F}'\, \hat{F} + 2(\hat{F}')^3)\hat{F}^{-3}| \\ &\leq (\beta_3 + 3\beta_2\beta_1 + 2\beta_1^3)(24/25)^{-3} \\ &\leq 6\beta_3(25/24)^3 \\ &\leq 7\beta_3\end{aligned}$$

(here $\beta_k = E|X|^k$, $k = 0,1,\ldots$).

Using now the inequality $|e^\alpha -1| \leq |\alpha|e^{|\alpha|}$ we obtain

$$\begin{aligned}|\hat{F}^n(tn^{-\frac{1}{2}})-e^{-t^2/2}| &= |\exp\ n \log \hat{F}(tn^{-\frac{1}{2}})-e^{-t^2/2}| \\ &\leq (7/6)|t|^3\beta_3 n^{-\frac{1}{2}} \exp\{ -t^2/2 + (7/6)|t|^3\beta_3\, n^{-\frac{1}{2}}\}.\end{aligned}$$

It remains only to notice that when t satisfies (18)

$$t^2/2 - (7/6)|t|^3\beta_3\, n^{-\frac{1}{2}} \geq t^2/4 \;. \square$$

<u>Proof 2 of the Berry-Esseen theorem</u> (the method of characteristic functions). As in the first proof it is enough to consider the case $\mu = 0, \sigma = 1$. By Corollary 2 we have for any $T > 0$

$$\sup_x |F_n(x)-\Phi(x)| \leq c[\int_{-T}^{T}|(\hat{F}^n(tn^{-\frac{1}{2}})-e^{-t^2/2})t^{-1}|dt + T^{-1}]$$

Take $T = n^{\frac{1}{2}}/(5\beta_3)$. Then by Lemma 2

$$\int_{-T_n}^{T_n} |(\hat{F}^n(tn^{-\frac{1}{2}})-e^{-t^2/2})t^{-1}|dt \leq (7/6)\beta_3 n^{-\frac{1}{2}} \int_{-\infty}^{\infty} t^2 e^{-t^2/4}\, dt = c\beta_3\, n^{-\frac{1}{2}} ,$$

which finishes the proof. $\square$

<u>Remark 1</u>. The speed $n^{-\frac{1}{2}}$ is "the true" speed of convergence in the central limit theorem . Indeed let $\{Y_j\}$ be a sequence of i.i.d. real random variables with $P(Y_1=0)= P(Y_1 = 1) = 1/2$. If $P_n(m) = P(\sum_1^n Y_i = m)$, then by the local DeMoivre-Laplace theorem

denoting $m_n = [n/2]$ (the integral part), we have

$$P_n(m_n)\sqrt{\pi/2}\, n^{\frac{1}{2}} \to 1 \text{ as } n \to \infty \quad . \tag{20}$$

On the other hand, denoting $X_i = Y_i - \frac{1}{2}$, we have $EX_i = 0, \sigma^2 = EX_i^2 = \frac{1}{4}$

$$P_n(m) = P(\bar{X}_n = 2(m-n/2)n^{-\frac{1}{2}}) \tag{21}$$

where $\bar{X}_n = \sigma^{-1}n^{-\frac{1}{2}} \sum_1^n X_i$. From (20), (21) it follows that for this sequence $\{X_i\}$

$$\sup_x |F_n(x)-\Phi(x)| \geq (1/2) \sup_x P(\bar{X}_n = x)$$

$$\geq (1/4)(2/\pi)^{\frac{1}{2}} n^{-\frac{1}{2}}$$

for all large enough n, which proves our assertion.

The classical results of Berry-Esseen type were extended and improved in several directions. Here are the main improvements:

1. One can construct estimates with less restrictions on the existence of moments. Let X_i be a sequence of i.i.d. real random variables with $EX_i = \mu, E(X_i-\mu)^2 = \sigma^2 < \infty$. Denote

$$\bar{\rho}_{3n} = \sigma^{-3} \int_{|x-\mu| \leq \sigma n^{\frac{1}{2}}} |x-\mu|^3 dF(x) + n^{\frac{1}{2}} \sigma^{-2} \int_{|x-\mu| > \sigma n^{\frac{1}{2}}} |x-\mu|^2 dF(x)$$

$$= \sigma^{-3} \int_{|x| \leq \sigma n^{\frac{1}{2}}} |x|^3 dF'(x) + n^{\frac{1}{2}} \sigma^{-2} \int_{|x| > \sigma n^{\frac{1}{2}}} |x|^2 dF'(x)$$

(F being the distribution function of X_1 and F' that of $X_1-\mu$).

Then one can show that (in our usual notations)

$$\sup_x |F_n(x)-\Phi(x)| \leq c\bar{\rho}_{3n}\, n^{-\frac{1}{2}} \quad .$$

Note that if $0 < \delta \leq 1$, then since $|x|^3 \leq |x|^{2+\delta} \sigma^{1-\delta} n^{(1-\delta)/2}$ when $|x| \leq \sigma n^{\frac{1}{2}}$ and $|x|^2 \leq |x|^{2+\delta} \sigma^{-\delta} n^{-\delta/2}$ when $|x| > \sigma n^{\frac{1}{2}}$, we have

$$\bar{\rho}_{3n} \leq n^{(1-\delta)/2} \sigma^{-2-\delta} \int |x|^{2+\delta} dx \; ,$$

i.e. $\bar{\rho}_{3n} n^{-\frac{1}{2}} \leq \beta_{2+\delta} n^{-\delta/2}$, where $\beta_{2+\delta} = E|X_1-m|^{2+\delta}$. Thus if $E|X_1|^{2+\delta} < \infty$, we obtain an estimate with the speed $n^{-\delta/2}$. One can show that in general the speed can not be better than $n^{-\delta/2}$ if we suppose only that $E|X_1|^{2+\delta} < \infty$

2. Under the same assumptions as in the Berry-Esseen theorem for any $x \in R$

$$|F_n(x) - \Phi(x)| \leq c(\beta_3/\sigma^3)n^{-\frac{1}{2}}(1+|x|^3)^{-1} \quad . \tag{22}$$

This estimate is called a non-uniform estimate (the name reflects the fact that both sides of inequality (22) depend on x) and obviously is better then the Berry-Esseen estimate when $|x|$ is large.

3. In the Berry-Esseen estimate the absolute moment can be replaced by the so-called pseudomoment $\nu = \int |x|^3 \, d|F''-\Phi|(x)$, where $F''(x) = P(X_1-\mu<\sigma x)$ and $|F''-\Phi|$ denotes the variation of $F''-\Phi$. More precisely, the following estimate is true

$$\sup_x |F_n(x)-\Phi(x)| \leq c \max(\nu, \nu^{\alpha(n)}) n^{-\frac{1}{2}}$$

where $\alpha(n) = \min(n/4,1)$. This estimate is better than Berry-Esseen one when the distribution of X_1 is close to the normal distribution in the sense that ν is small (recall that $\beta_3/\sigma^3 \geq 1$ always). Note that $\alpha(n) = \min(n/4,1)$ is unimprovable here.

4. For any real function f on R define

$$\omega_f(R) = \sup\{|f(x)-f(y)|, x,y \in R\}$$

$$\omega_f(x,\varepsilon) = \sup\{|f(y)-f(z)|, |y-x| < \varepsilon, |z-x| < \varepsilon\}$$

$$f_y(x) = f(x+y).$$

If f is Borel measurable and G is a distribution function on R, denote

$$\bar{\omega}_f(\varepsilon,G) = \int \omega_f(x,\varepsilon) dG(x)$$

$$\omega_f^*(\varepsilon,G) = \sup\{\bar{\omega}_{f_y}(\varepsilon,G), y \in R\}$$

If now f is a bounded Borel measurable function on R, then (supposing $\sigma = 1$ for simplicity)

$$|\int f \, d(F_n-\Phi)(x)| \leq c[\omega_f(R)\varepsilon_n + \omega_f^*(c_1 \varepsilon_n, \Phi)] \tag{23}$$

where $\varepsilon_n = \beta_3 n^{-\frac{1}{2}}$. This is indeed a generalization of the Berry-Esseen theorem since if we take f: $f(z)=1$ for $z< x$, $= 0$ for $z \geq x$, then the left side in (23) is $|F_n(x)-\Phi(x)|$, $\omega_f(R) = 1$ and, as it is easy to check

$$\omega_{f_y}(z,\varepsilon) = \begin{cases} 0 \text{ when } z \leq x-y-\varepsilon \text{ or } z \geq x-y+\varepsilon \\ 1 \text{ otherwise} \end{cases}$$

$$\bar{\omega}_{f_y}(\varepsilon,\Phi) = \int \omega_{f_y}(z,\varepsilon) d\Phi(x) \leq (2/\pi)^{\frac{1}{2}}\varepsilon ;$$

thus the right side in (23) is majorised by $\varepsilon_n = c\beta_3 n^{-\frac{1}{2}}$.

The estimate (23) is a rather general estimate which makes sense for an arbitrary bounded Borel measurable function f and at the same time it is precise (it implies the Berry-Esseen estimate). The right side of (23) depends essentially on f and this is something one should expect since in general we have only weak convergence of F_n to Φ.

5. The summands X_j may not be identically distributed. Let $X_1, X_2, \ldots$ be independent real random variables and let $EX_j = \mu$, $E(X_j-\mu_j)^2 = \sigma_j^2, j=1,2,\ldots$. Denote $s_n^2 = \sum_1^k \sigma_j^2$. If $s_n \neq 0$ then

$$|P(s_n^{-1} \sum_1^n (X_i-\mu_i) < x) - \Phi(x)| \leq c s_n^{-3} \sum_{i=1}^n E|X_i-\mu_i|^3$$

This estimate was obtained already by C.-G. Esseen himself.

6. The Berry-Esseen estimate has been generalized to the multidimensional case. Let $\{X_i = (X_{i1},\ldots,X_{ik}), i = 1,2,\ldots\}$ be i.i.d. random variables with values in R^k. Denote $EX_1 = \mu$ and let V be the covariance matrix of X_1 with elements $v_{ij} = E(X_{1i}-\mu_i)\cdot(X_{1j}-\mu_j)$. Denote also by P_n the distribution of the normalized sum $n^{-\frac{1}{2}} \sum_1^n V^{-\frac{1}{2}}(X_i-\mu)$. Recall that N is the standard normal distribution in R^k and C is the class of all measurable convex sets in R^k. A multidimensional version of the Berry-Esseen theorem states that there exists a constant $c(k)$ such that for all $n = 1,2,\ldots$

$$\sup_{A \in C} |P_n(A)-N(A)| \leq c(k)\rho_3 n^{-\frac{1}{2}}$$

where $\rho_3 = E|V^{-\frac{1}{2}}(X_1-\mu)|^3$.

Remark 2. Since, as it is easy to see, for any two real random variables X,Y

$$|P(X<x) - P(Y<x)| \leq \sup_{A \in C} |P(X \in A) - P(Y \in A)|$$
$$\leq 2\sup_x P(X<x) - p(Y<x)|.$$

in the one-dimensional case the problem of estimation of $\sup\{|F_n(x)-\Phi(x)|, x \in R\}$ is equivalent to the problem of estimation of $\sup\{|P_n(A)-N(A)|, A \in C\}$ (up to a constant).

We have described the main directions of generalizations and improvements of the Berry-Esseen estimate by indicating a simple result in each of the directions. Of course a number of estimates has been constructed which combine in itself several of the above extentions and improvements. We have chosen to separate them to be able to make clear the main ideas behind general estimates. Some general estimates will appear in what follows .

§ 2. The method of compositions. Estimates for convex and arbitrary Borel sets.

In this section we shall obtain some estimates of the speed of convergence in the multidimensional central limit theorem using the method of compositions. Estimates will be given for deviations on convex sets and (under some restrictions) arbitrary Borel sets. To make the proofs somewhat more simple and transparent we assume the existence of the third moment (the estimates become trivial when it does not exist). Note that this assumption can be considerably weakened.

Theorem 1. Let $X_1, X_2, \ldots$ be a sequence of independent random variables with values in R^k with the same distribution P. Suppose that $E|X_1|^2 < \infty$ and that the covariance matrix V of P is nondegenerate. Denote P_n the distribution of the normalized sum $n^{-\frac{1}{2}} \sum_1^n V^{-\frac{1}{2}} (X_i-\mu)$, where $\mu=EX_1$, and let N be the standard normal distribution in R^k. Then

1. There exist $\bar{c}(k)$ such that for all $n = 1,2, \ldots$

$$\sup_{A \in C} |P_n(A)-N(A)| \leq \bar{c}(k)\hat{\nu}n^{-\frac{1}{2}} \tag{1}$$

where

$$\hat{\nu} = \begin{cases} \nu \text{ if } \nu \geq 1, \\ \max(\nu, \nu^{nk/(k+3)}) \text{ if } \nu < 1, \end{cases} \qquad \nu = \int |x|^3 |\bar{P} - N| (dx) \tag{2}$$

and $\bar{P}$ is the distribution of $V^{-\frac{1}{2}}(X_1-\mu)$ (ν is called 3rd preudomoment of $\bar{P}$).

2. There exist two constants c_1, c_2 such that if $\nu < c_1$ then for all $n = 1,2,\ldots$

$$\delta_n = |P_n - N| (R^k) \leq c_2 \hat{\nu} n^{-\frac{1}{2}} . \tag{3}$$

The constant $\bar{c}(k)$ may be taken to be

$$\bar{c}(k) \sim \bar{c}k. \tag{4}$$

Remark 1. Note that in the definition (2) of $\hat{\nu}$ the power nk/(k+3) is the best possible, i.e. it cannot be replaced by a larger one. Indeed, consider the Borel probability measure P_ε, $\varepsilon > 0$, on R^k defined by

$$P_\varepsilon(B) = N(B \cap S_\varepsilon^c) + (2k)^{-1} N(S_\varepsilon) \sum_{j=1}^k (\chi_B(s\varepsilon e_j)+\chi_B(-s\varepsilon e_j))$$

where e_j are the coordinate unite vectors and

$$s = \varepsilon^{-2}(N(S_\varepsilon))^{-1} \int_{S_\varepsilon} |x|^2 N(dx) .$$

Note that P_ε has mean 0 and covariance matrix I. For this measure P_ε we have

$$\begin{aligned} \nu = \nu_\varepsilon &= \int |x|^3 \, |P_\varepsilon - N|(dx) \\ &= \varepsilon^3 N(S_\varepsilon) + \int_{S_\varepsilon} |x|^3 N(dx) \\ &\leq c(k)\varepsilon^{k+3} . \end{aligned}$$

On the other hand for small ε

$$\begin{aligned} \Delta_n = \Delta_n(\varepsilon) &= \sup_{A \in C} |P_n(A) - N(A)| \\ &\geq P_\varepsilon(\{s\varepsilon e_j\})^n \\ &= (2k)^{-n}(N(S_\varepsilon))^n \\ &\geq (c_1(k))^n \varepsilon^{nk}. \end{aligned}$$

Thus for small ε

$$\Delta_n(\varepsilon) \geq c(k,n)\nu_\varepsilon^{nk/(k+3)}$$

and $\nu_\varepsilon \to 0$ as $\varepsilon \to 0$. This proves our assertion.

To prove Theorem 1 we shall need several lemmas.

Lemma 1. If Q is any probability distribution on R^k then

$$|Q-N|(R^k) \leq ck^{-3/2} \nu_1^{k/(k+3)} \tag{5}$$

where $\nu_1 = \int |x|^3 \, |Q-N|(dx)$. If moreover Q has mean zero and covariance matrix I then for all $n \geq 1$

$$|Q_n - N|(R^k) \leq [|Q-N|(R^k)]^n + c(n)\nu_1 , \tag{6}$$

where Q_n is defined by $Q_n(B) = Q^{*n}(n^{\frac{1}{2}}B)$.

Proof. To prove (5) suppose first that $0 < v = |Q-N|(R^k) < 2$. Let $R^k = R^+ \cup R^-$ be the Hahn decomposition of R^k with respect to Q-N, i.e. R^+, R^- are Borel sets such that $R^+ \cap R^- = \emptyset$, $(Q-N)(B) \geq 0$ if $B \subset R^+$, $(Q-N)(B) \leq 0$ if $B \subset R^-$. Define measure M by

$$M(B) = N(B \cap R^+) + Q(B \cap R^-) .$$

We have

$$\int |x|^3 \, |Q-N|(dx) \ \geq \ \int |x|^3 \, |M-N|(dx)$$

$$= \int |x|^3 \, (N-M)(dx) \qquad (7)$$

Note that $M(R^+) = 1 - v/2$. Indeed

$$(Q-N)(R^+) + (Q-N)(R^-) = 0$$

$$(Q-N)(R^+) - (Q-N)(R^-) = v \qquad (8)$$

thus $(Q-N)(R^+) = v/2$, and it follows that

$$M(R^k) = N(R^+) + Q(R^-)$$

$$= N(R^+) + 1-Q(R^+)$$

$$= 1-v/2 \ .$$

Take positive number r such that $N(S_r) = v/2$. Then

$$N(S_r^c) = 1-v/2 = M(S_r) + M(S_r^c)$$

and

$$M(S_r) = (N-M)(S_r^c).$$

Thus we have

$$\int_{S_r} |x|^3 \, M(dx) \ \leq \ r^3 \, M(S_r)$$

$$= r^3 (N-M)(S_r^c)$$

$$\leq \int_{S_r^c} |x|^3 \, (N-M)(dx)$$

and it follows that

$$\int |x|^3 (N-M)(dx) = \left(\int_{S_r} + \int_{S_r^c} \right) |x|^3 (N-M)(dx)$$

$$\geq \int_{S_r} |x|^3 (N-M)(dx) + \int_{S_r} |x|^3 \, M(dx)$$

$$= \int_{S_r} |x|^3 N(dx) \ . \qquad (9)$$

Now (7) and (9) imply

$$\nu_1 = \int |x|^3 \, |Q-N|(dx) \geq \int_{S_r} |x|^3 \, N(dx)$$

and we have, since $|Q-N|(R^k) = v = 2N(S_r)$,

$$\frac{|Q-N|(R^k)}{\nu_1^{k/(k+3)}} \leq \frac{2N(S_r)}{\left(\int_{S_r} |x|^3 \, N(dx)\right)^{k/(k+3)}}$$

$$= f(r,k) \, .$$

It is easy to check that $f(r,k)$ as a function of r is nondecreasing and $f(+0,k) \leq ck^{-3/2}$, which proves (5) when $v \neq 0,2$.

If $v = 0$, (5) is obvious, and if $v = 2$, then Q and N are orthogonal ((8) implies in this case $(Q-N)(R^+) = 1$, and thus $Q(R^+) = 1$, $N(R^+) = 0$), and we have

$$\left(\int |x|^3 |Q-N|(dx)\right)^{k/(k+3)} \geq \left(\int |x|^3 N(dx)\right)^{k/(k+3)} = ck^{3/2} |Q-N|(R^k) \, .$$

Inequality (5) is proved.

To prove (6) consider first the case $n = 2$. Noting that $N(.) = N^n(n^{\frac{1}{2}}.)$ for all n, we have for any Borel set B

$$\begin{aligned} |Q_2(B) - N(B)| &= |Q^2(\sqrt{2}\,B) - N^2(\sqrt{2}\,B)| \\ &= |(Q-N)^2(\sqrt{2}\,B) + 2(Q-N) * N(\sqrt{2}\,B)| \, . \end{aligned} \tag{10}$$

Now since $(Q-N)(E) \leq (1/2)|Q-N|(R^k)$ for any Borel set E, we have

$$|(Q-N)^2(\sqrt{2}\,B)| \leq (1/2)|Q-N|(R^k) \int |Q-N|(dx) = (1/2)[|Q-N|(R^k)]^2 \, . \tag{11}$$

Furthermore if $g(x)$ is a bounded Borel measurable function then for any $t > 0$

$$\psi(x) = \int g(x+y) N_t(dy) = \int g(z) \phi_t(x-z)\,dz$$

is a smooth function and by Taylor's formula

$$\psi(x) = \sum_{\|a\| \leq 2} \frac{x^a}{a!} D_a \psi(0) + \sum_{\|a\| = 3} \frac{x^a}{a!} D_a \psi(\theta x) \tag{12}$$

where a are nonnegative integral vectors and $|\theta| \leq 1$. For $s > 0$ and Borel sets E denoting $H_{(s)} = Q(sE) - N(sE)$ and observing that $H_{(s)}$ has the first and second moments equal to zero we have

$$\begin{aligned} \left|\int g(x) N_t * H_{(s)}(dx)\right| &= \left|\int \psi(x) H_{(s)}(dx)\right| \\ &\leq \int \Big| \sum_{\|a\| = 3} \frac{x^a}{a!} D_a \psi(\theta x) \Big| \, |H_{(s)}|(dx) \, . \end{aligned}$$

But for any y

$$\sum_{\|a\|=3} \frac{x^a}{a!} D_a\psi(y) = \frac{t^6|x|^3}{6} \int g(z)\Big(\sum_{u=1}^{k} (z_u - y_u)x_u/|x|\Big)^3 \phi_t(z-y)\,dz$$
$$- \frac{t^4|x|^3}{2} \int g(z)\Big(\sum_{u=1}^{k} (z_u - y_u)x_u/|x|\Big)\phi_t(z-y)\,dz\,, \qquad (13)$$

where x_u (resp. y_u, z_u) are the coordinates of x (resp. y,z) and hence, since ϕ is symmetric

$$\Big|\sum_{\|a\|=3} \frac{x^a}{a!} D_a\psi(y)\Big| \le \frac{|x|^3}{2} \sup_z |g(z)|\Big(\frac{t^6}{3}\int |z_1|^3\phi_t(z)\,dz + t^4\int |z_1|\phi_t(z)\,dz\Big)$$
$$\le c \sup_z |g(z)|\, t^3|x|^3 \,.$$

Thus, since $\int |x|^3 |H_{(s)}|(dx) = s^{-3}\nu_1$,

$$\Big|\int g(x) N_t * H_{(s)}(dx)\Big| \le c \sup_z |g(z)|\, s^{-3}t^3\nu_1 \,. \qquad (14)$$

From (10), (11) and (14) with $g(x) = \chi_{\sqrt{2}B}(x)$ and $s,t=1$ we now deduce (6) for $n=2$.

When $n>2$ let us observe at first that for any Borel set B

$$|(Q-N)^n(B)| \le (1/2)[|Q-N|(R^k)]^n, \quad n = 1,2,\dots \,.$$

Indeed we have seen that this is true for $n=1,2$ and for $n>2$ using induction on n we have

$$|(Q-N)^n(B)| = \Big|\int (Q-N)^{n-1}(B-x)(Q-N)(dx)\Big|$$
$$\le \sup_E |(Q-N)^{n-1}(E)| \int |Q-N|(dx)$$
$$\le (1/2)[|Q-N|(R^k)]^n$$

(the sup here is taken over all Borel sets E). Now, using the above estimates, we obtain

$$|Q_n(B) - N(B)| = |Q^n(n^{\frac12}B) - N^n(n^{\frac12}B)|$$
$$= \Big|\sum_{i=1}^{n} \binom{n}{i}(Q-N)^i * N^{n-i}(n^{\frac12}B)\Big|$$
$$\le |(Q-N)^n(n^{\frac12}B)|$$
$$+ \Big|\int (Q-N) * N(n^{\frac12}B - x)\Big(\sum_{i=1}^{n-1}\binom{n}{i}(Q-N)^{i-1} * N^{n-i-1}\ (dx)\Big|$$
$$\le (1/2)[|Q-N|(R^k)]^n + \sup_E |(Q-N) * N(E)|\cdot c(n)$$
$$\le (1/2)[|Q-N|(R^k)]^n + c(n)\nu_1 \,. \quad \square$$

In the proof of the next lemma we shall need one fact from the theory of convex sets in R^k. Recall that a convex subset of R^k is called regular if all its boundary points and all its support hyperplanes are regular. By definition a support hyperplane of a convex set is regular if it meets its boundary in only one point and a boundary point of a convex set is regular if it lies on only one support hyperplane. The result we shall need is: if A is a bounded convex set with non-empty interior then for any $\delta > 0$ there exist two bounded closed regular convex sets R_1, R_2 such that

$$R_1 \subset A \subset R_2, \quad R_2 \subset R_1^{\delta}. \tag{15}$$

Regarding this result one can consult, e.g. [13].

Recall also that for any set E by ∂E we denote its boundary and by E^{ε} its ε-neighbourhood, i.e. $E^{\varepsilon} = \cup\{S_{\varepsilon}(x), x \in E\}$, $S_{\varepsilon}(x)$ being the open ball of radius ε with center at x.

Lemma 2. There exist a constant $c(k,s), s \geq 0$, such that for any $\varepsilon > 0$ and any convex $A \subset R^k$,

$$\int_{(\partial A)^{\varepsilon}} |x|^s N(dx) < c(k,s)\varepsilon.$$

Proof. We shall prove the lemma for the case $k = 2$. With trivial changes the proof carries over to the case of any $k \geq 2$. The set $(\partial A)^{\varepsilon}$ may be represented as

$$(\partial A)^{\varepsilon} = A^{\varepsilon} \setminus A^{-\varepsilon} = (A^{\varepsilon} \setminus A) \cup (A \setminus A^{-\varepsilon})$$

where $A^{-\varepsilon} = \cup\{x : S_{\varepsilon}(x) \subset A\}$. We shall show that

$$\int_{A^{\varepsilon} \setminus A} |x|^s N(dx) \leq c(s)\varepsilon. \tag{16}$$

The same bound for the integral over the set $A \setminus A^{-\varepsilon}$ can be obtained in a similar way.

It is enough to consider only bounded convex sets since for any $r > 0$

$$A^{\varepsilon} \setminus A \subset ((A \cap S_{r+\varepsilon})^{\varepsilon} \setminus (A \cap S_{r+\varepsilon})) \cup S_r^c$$

and $\int_{S_r^c} |x|^s N(dx) \to 0$ as $r \to \infty$. Now, if A has an empty enterior, then A is contained in a line L, say, and if u is the distance from the origine to L, then

$$\int_{A^{\varepsilon} \setminus A} |x|^s N(dx) \leq \int_{L^{\varepsilon}} |x|^s N(dx)$$
$$\leq c(s) \int_{x_1 = u-\varepsilon}^{u+\varepsilon} (|x_1|^s + |x_2|^s) N(dx)$$
$$\leq c(s)\varepsilon$$

(here x_1, x_2 are the coordinates of x in a system of coordinates with the 1st axis orthogonal to L).

Thus we can suppose that A is bounded and with interior points. Then for any $\delta > 0$ there exist regular bounded convex closed sets R_1, R_2 satisfying (15). If the lemma is true for such regular convex sets, then, since obviously $A^{\varepsilon} \setminus A \subset R_1^{\delta+\varepsilon} \setminus R_1$, we have

$$\int_{A^{\varepsilon} \setminus A} |x|^s \, N(dx) \leq \int_{R_1^{\delta+\varepsilon} \setminus R_1} |x|^s \, N(dx) \leq c(s)(\delta+\varepsilon)$$

for any $\delta > 0$, and hence the lemma is true for any convex set.

Now let A be a regular closed bounded convex set. Denote by $A_1(A_2)$ the set of those points of ∂A at which the cosine of the angle between the external normal to A and the positive (negative) direction of the 1st axis is $\geq 2^{-\frac{1}{2}}$. The sets A_3, A_4 are defined similarly with respect to the 2nd axis. Clearly $\partial A = \bigcup_1^4 A_i$. Denote now $A_i' = \cup\{n_x^{\varepsilon},\ x \in A_i\}$, n_x^{ε} being the set of points on the external normal to A at x with distance from x not greater then ε. The sets A_i' are obviously bounded and closed and we have

$$A^{\varepsilon} \setminus A \subset \bigcup_1^4 A_i' \tag{17}$$

Indeed let $y \in A^{\varepsilon} \setminus A$ and let y' be the nearest point in ∂A to y. We claim that $y \in n_{y'}^{\varepsilon}$. If it is not so consider the line ℓ which is perpendicular to y'y and contains y'. There exists a point z of A belonging to the same open half plane defined by ℓ as y. The segment joining z and y' will contain then a point z' of A which is closer to y than y'. The obtained contradiction proves (17).

Let now L be a straight line which is parallel to the 1st axis and interesects A_1'. Denote by x_0' the point in $L \cap A_1'$ at which the 1st coordinate of the points of the set $L \cap A_1'$ attains its minimum. Let $x_0 \in A_1$ be the point such that $x_0' \in n_{x_0}^{\varepsilon}$ and x_0'' be the point on the external normal to A at x_0 with the distance to x_0 equal to ε. Draw the line ℓ through the point x_0'' which is parallel to the support line to A at x_0.

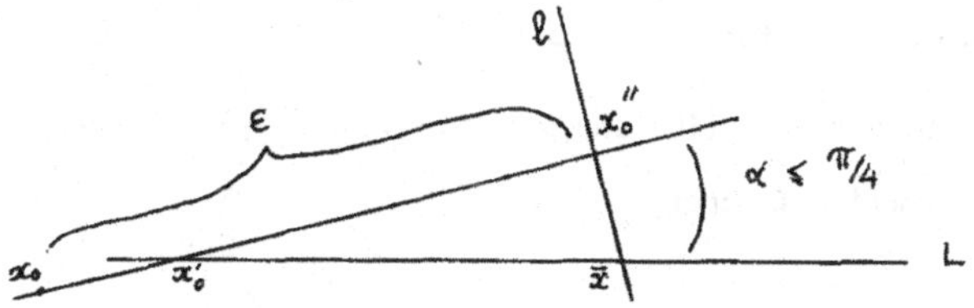

It follows from the definition of A_1' that ℓ intersects L in some point $\bar{x}$, whose first co-ordinate is greater than the first coordinate of x_0' and whose distance from x_0' does not exceed $\varepsilon\sqrt{2}$. Since the set A^ε lies on the same side of the line ℓ as the point x_0', the set $L \cap A_1'$ is contained in the segment joining x_0' and $\bar{x}$ (of length $\leq \varepsilon\sqrt{2}$). In other words, we have shown that the section $A_1'(y_2)$ of the set A_1' by any straight line of the form $\{x = (x_1,x_2) : x_2 = y_2\}$ can be covered by a segment of length $\leq \varepsilon\sqrt{2}$. From this we have

$$\int_{A_1'} |x|^s N(dx) = \frac{1}{2\pi} \int_{A_1'} |x|^s \exp\{-(1/2)(x_1^2+x_2^2)\}dx_1\, dx_2$$

$$\leq \frac{c(s)}{2\pi} \int_{A_1'} e^{-\frac{1}{2}x_2^2} \Big(\int_{A_1'(x_2)} |x_1|^2 + |x_2|^2)e^{-\frac{1}{2}x_1^2} dx_1\Big)dx_2$$

$$\leq c(s)\varepsilon.$$

Similarly $\int_{A_i'} |x|^s N(dx) \leq c(s)\varepsilon$, $i = 2,3,4$. In view of (17) this proves the lemma. □

Remark 2. Later we shall need estimates for $c(k,0)$ and $c(k,3)$ in Lemma 4. From the k-dimensional version of the given proof of Lemma 2 it follows that $c(k,0) \leq ck^{3/2}$, $c(k,3) \leq ck^{5/2}$ (see [32], Lemma 1). A considerably more complicated proof (see e.g. [8], Theorem 3.1) leads to better estimates $c(k,0) \leq ck^{\frac{1}{2}}$, $c(k,3) \leq ck^2$. On the other hand it can be shown that $c(k) \geq c_1(\log k)^{c_2}$ (see [11]). The question of the true asymptotic behaviour of $c(k,s)$ (in particular of $c(k,0)$) is still open.

For any real function f on R^k denote

$$\hat{f}^\varepsilon(x) = \sup\{f(y), y \in S_\varepsilon(x)\},\quad \check{f}^\varepsilon(x) = \inf\{f(y), y \in S_\varepsilon(x)\}.$$

Since $\{x: \hat{f}^\varepsilon(x) > u\} = \cup\{S_\varepsilon(y): f(y) > u\}$, the functions $\hat{f}^\varepsilon(x)$ and $\check{f}^\varepsilon(x) = -\widehat{(-f)}^\varepsilon$ are Borel measurable.

Lemma 3. Let P,Q,R be any three probability measures on R^k. Assume that

$$R(S_v) \geq u \quad \text{for some } v > 0,\ 2^{-1} < u \leq 1 . \tag{18}$$

Define measure R_T by $R_T(E) = R(TE)$ and put $t = vT^{-1}$, $f_x(y) = f(x+y)$. For any bounded Borel measurable function f denote

$$\delta = \sup_x \{|\int f_x(y)H(dy)|\},\ H = P - Q$$

$$\delta_T = \sup_x \{\max[|\int \hat{f}_x^t(y)H*R_T(dy)|, |\int \check{f}_x^t(y)H*R_T(dy)|]\}$$

$$\gamma_T = \sup_x \{\max[|\int \hat{f}_x^{2t}(y)-f_x(y))Q(dy), \int(f_x(y)-\check{f}_x^{2t}(y))Q(dy)]\}.$$

Then for any $T > 0$

$$\delta \leq (2u-1)^{-1} (\delta_T+\gamma_T).$$

Proof. Suppose that $\delta = \sup\{\int f_x(y)H(dy)\}$. The case when $\delta = -\inf\{\int f_x(y)H(dy)\}$ is treated in a similar way. For any $\varepsilon > 0$ there exist x_ε such that $\int f_{x_\varepsilon}(y)H(dy) > \delta - \varepsilon$. Denoting $f_1 = f_{x_\varepsilon}$ we have

$$\delta_T \geq (\int_{S_t} + \int_{S_t^c}) [\int \hat{f}_1^t(y+z)H(dz)R_T(dy) .$$

If $|y| < t$

$$f_1(z) \leq \hat{f}_1^t(y+z) \leq \hat{f}_1^{2t}(z)$$

and

$$\begin{aligned}\int \hat{f}_1^t(y+z)H(dz) &\geq \int f_1(z)(dz) + \int(f_1(z)-\hat{f}_1^t(z+y))Q(dz)\\ &\geq \delta - \varepsilon - \int(\hat{f}_1^{2t}(z)-f_1(z))Q(dz)\\ &\geq \delta - \varepsilon - \gamma_T .\end{aligned}$$

On the other hand for any $y \in R^k$

$$\begin{aligned}\int \hat{f}_1^t(z+y)H(dz) &\geq \int f_1(z+y)H(dz) + \int(f_1(z+y)-\hat{f}_1^t(z+y))Q(dz)\\ &\geq - \delta - \int(\hat{f}^t_{x_\varepsilon+y}(z) - f_{x_\varepsilon+y}(z))Q(dz)\\ &\geq - \delta - \gamma_T ,\end{aligned}$$

since, as it easily follows from the definitions, $f_1(z+y) = f_{x_\varepsilon+y}(z)$ and $\hat{f}_1^t(z+y) = \hat{f}^t_{x_\varepsilon+y}(z)$. Thus

$$\begin{aligned}\delta_T &\geq (\delta-\varepsilon-\gamma_T)R_T(S_t) - (\delta+\gamma_T)R_T(S_t^c)\\ &\geq \delta(2R_T(S_t)-1)-\gamma_T - \varepsilon.\end{aligned}$$

Since ε is arbitrary small and, by (18), $R_T(S_t) = R(S_v) \geq u$, this proves the lemma.□

Corollary 1. Preserving the notations and assumptions of Lemma 3 for a Borel set B define

$$\delta' = \sup_x |H(B+x)|$$

$$\delta_T' = \sup_x \{\max[|(H*R_T)(B^t+x)|, |(H*R_T)(B^{-t}+x)|]\}$$

$$\gamma_T' = \sup_x \{\max[R((B^{2t}\setminus B)+x), R((B\setminus B^{-2t})+x)]\}.$$

Then

$$\delta' \leq (2u-1)^{-1}(\delta_T' + \gamma_T').$$

Proof. Take $f(x) = \chi_B(x)$, the indicator function of the set B. Then the definitions imply easily that

$$f_x(y) = \chi_{B-x}(y)$$

$$\hat{f}_x^t(y) = \chi_{B^t-x}(y), \quad \check{f}_x^t(y) = \chi_{B^{-t}-x}(y)$$

$$\hat{f}_x^{2t}(y)-f_x(y) = \chi_{(B^{2t}\setminus B)-x}(y), \quad f_x(y)-\check{f}_x^{2t}(y) = \chi_{(B\setminus B^{-2t})-x}(y).$$

Thus for this function f, $\delta=\delta'$, $\delta_T = \delta_T'$, $\gamma_T = \gamma_T'$ and Lemma 3 implies the Corollary.□

Corollary 2. For any probability measure P on R^k

$$\sup_{A\in\mathcal{C}} |(P-N)(A)| \leq 2[\sup_{A\in\mathcal{C}}|(P-N)*N_T(A)| + ckT^{-1}].$$

Proof. Take in Corollary 1, Q = R = N and observe that if A is convex then both A^t and A^{-t} are convex. Note also that

$$(A^{2t}\setminus A)+x = (A+x)^{2t}\setminus (A+x)$$

$$(A\setminus A^{-2t})+x = (A+x)\setminus (A+x)^{-2t}.$$

Thus Corollary 1 with u = 3/4 and Lemma 2 imply

$$\sup_{A\in\mathcal{C}}|(P-N)(A)| \leq 2[\sup_{A\in\mathcal{C}} |(P-N)*N_T(A)| + c(k)2vT^{-1}],$$

where $c(k)$ is the same as in Lemma 2. According to Remark 2 we may take $c(k) \leq ck^{\frac{1}{2}}$ On the other hand for v we have the relation $N(S_v) \geq u = 3/4$. Take $v = v(k)$ such that $N(S_v) = 3/4$. To finish the proof of the Corollary it is enough to show now that $v(k)k^{-\frac{1}{2}} \to 1$ as $k \to \infty$.

We shall show that, more generally, if $0 < w < 1$ and $v = v(k)$ is such that $N(S_v) = w$, then

$$v(k)k^{-1} \to 1 \quad \text{as} \quad k \to \infty \quad . \tag{19}$$

Let $Z_1, Z_2, \ldots$ be i.i.d. normal (0,1) random variables. Applying the central limit theorem we have for any $\lambda > 0$

$$N(S_{\lambda k^{\frac{1}{2}}}) = P\left(\left(\sum_1^k Z_j^2\right)^{\frac{1}{2}} < \lambda k^{\frac{1}{2}}\right)$$

$$= P\left(k^{-\frac{1}{2}}\left(\sum_1^k (Z_j^2-1)\right) < (\lambda^2-1)k^{\frac{1}{2}}\right)$$

$$\to \begin{cases} 1, & \lambda > 1 \\ \frac{1}{2}, & \lambda = 1 \\ 0, & 0 < \lambda < 1. \end{cases}$$

Since $0 < P((\sum_1^k Z_j^2)^{\frac{1}{2}} < v(k)) = w < 1$, for any λ_1, λ_2 such that $0 < \lambda_1 < 1 < \lambda_2$ we thus have

$$\lambda_1 k^{\frac{1}{2}} < v(k) < \lambda_2 k^{\frac{1}{2}}$$

when k is large enough. Hence $v(k)k^{-\frac{1}{2}} \to 1$ as $k \to \infty$ which finishes the proof. □

For any real function f on R^k, $E \subset R^k$, $x,y \in R^k, \varepsilon > 0$ define

$$\omega_f(E) = \sup\{|f(z_1) - f(z_2)|,\ z_1, z_2 \in E\}$$

$$\omega_f(x,\varepsilon) = \omega_f(S_\varepsilon(x)) \tag{20}$$

$$f_y(x) = f(x+y).$$

If f is Borel measurable and M is a Borel measure on R^k, denote

$$\bar{\omega}_f(\varepsilon,M) = \int \omega_f(x,\varepsilon)M(dx)$$

$$\omega_f^*(x,M) = \sup\{\bar{\omega}_{f_y}(\varepsilon,M), y \in R^k\} \tag{21}$$

<u>Corollary 3</u>. Preserving the notations and assumptions of Lemma 3 we have

$$\left|\int f(x)(P-Q)(dx)\right| \le (2u-1)^{-1}[\omega_f(R^k)|(P-Q)*R_T|(R^k) + \omega_f^*(2vT^{-1},Q)].$$

<u>Proof</u>. We have

$$\left|\int f(x)(P-Q)(dx)\right| \le \delta\ ,\ \delta_T \le \omega_f(R^k)|(P-Q)*R_T|(R^k)$$

and

$$\gamma_T \le \sup_x \int(\hat{f}_x^{2t}(y) - \check{f}_x^{2t}(y))Q(dy)$$

$$= \omega_f^*(2t,Q)$$

since $\hat{f}_x^{2t}(y) - \check{f}_x^{2t}(y) = \omega_{f_x}(y,2t)$, and the Corollary follows. □

Proof of Theorem 1. Without loss of generality we may suppose that $\mu = 0$, $V = I$. The general case is reduced to this one by considering the sequence of random variables $V^{-\frac{1}{2}}(X_j-\mu), j=1,2,\ldots$ which have mean 0 and covariance matrix I. We shall assume also that $\nu < \infty$, or, equivalently, that $E|X_1|^3 < \infty$ (otherwise the theorem is trivial). If

$$n \leq (k+3)k^{-1} \tag{22}$$

then since P has mean 0 and covariance matrix I, we have by Lemma 1

$$\begin{aligned}\sup_{A\in\mathcal{C}} |P_n(A)-N(A)| &\leq (1/2)\ |P_n-N|(R^k)\\ &\leq c^n\, k^{-3n/2}\, \nu^{kn/(k+3)} + c(n)\nu\\ &\leq c\,\hat{\nu}\, n^{-\frac{1}{2}}\end{aligned}$$

(note that the condition (22) implies $n \leq 4$). This proves the theorem in the case (22).

Let now $n > (k+3)k^{-1}$. In this case we have $\hat{\nu} = \nu$. We may also suppose that

$$\nu \geq c_1 \tag{23}$$

(when (23) is violated we have (3) and (1) follows from (3)).

To prove (1) we shall show now that if it is true for all values of n smaller than some fixed value, with constant $\bar{c}(k)$ large enough to satisfy (4), then it is also true for this fixed value of n with the same constant $\bar{c}(k)$.

Denote by $P_{(n)}$ the distribution of $n^{-\frac{1}{2}} X_1$. Since $P_n = P^n_{(n)}$ and $N = N^n_{n^{\frac{1}{2}}}$, we have for any $T > 0$ (cf. § 1, (10))

$$\begin{aligned}(P_n-N)*N_T &= (P^n_{(n)} - N^n_{n^{\frac{1}{2}}})*N_T\\ &= (P_{(n)}-N_{n^{\frac{1}{2}}})*\Big(\sum_{i=0}^{n-1} P^i_{(n)}*N^{n-i-1}_{n^{\frac{1}{2}}}\Big)*N_T\\ &= \Big[\sum_{i=1}^{n-1} (P^i_{(n)}-N^i_{n^{\frac{1}{2}}})*N^{n-i-1}_{n^{\frac{1}{2}}} * N_T + nN^{n-1}_{n^{\frac{1}{2}}} * N_T\Big]*(P_{(n)}-N_{n^{\frac{1}{2}}})\\ &= \Big(\sum_{i=1}^{n-1} U_i + nU_0\Big)*H_1\,,\end{aligned} \tag{24}$$

where

$$U_i = H_i * N_{\tau_i}, \quad i = \overline{0,n-1}, \quad H_i = P^i_{(n)} - N^i_{n^{\frac{1}{2}}}, \quad i = \overline{1,n-1}$$

$$H_0 = \delta_0, \quad \tau_i = ((n-i-1)/n + T^{-2})^{-\frac{1}{2}}, \quad i = \overline{0,n-1} \tag{25}$$

(δ_0 is the probability distribution concentrated at the origin).

Fix a convex set A. To estimate

$$|U_i * H_1(A)| = \int H_i(A-x) N_{\tau_i} * H_1(dx)$$

we can apply (14) with

$$g(x) = H_i(A-x), \quad t = \tau_i, \quad s = n^{\frac{1}{2}}$$

(note that $H_1 = H_{(n^{\frac{1}{2}})}$ in the notation of (14)). This gives

$$|U_i * H_1(A)| \le c \sup_x |H_i(A-x)| n^{-3/2} \tau_i^3 \, \nu \, . \tag{26}$$

Moreover, for $i \ge (k+3)k^{-1}$ we have by the inductive hypothesis for any convex Borel set B

$$\begin{aligned} |H_i(B)| &= |P^i_{(n)}(B) - N^i_{n^{\frac{1}{2}}}(B)| \\ &= |P_i((n/i)^{\frac{1}{2}}B) - N((n/i)^{\frac{1}{2}}B)| \\ &\le \bar{c}(k)\nu i^{-\frac{1}{2}} \, . \end{aligned} \tag{27}$$

If $i < (k+3)k^{-1}$ we shall use the obvious estimate

$$|H_i(x)| \le 1 \, . \tag{28}$$

From (26) - (28) it follows now that

$$|U_i * H_1(A)| \le \begin{cases} \bar{c}(k) c \nu^2 \tau_i^3 \, i^{-\frac{1}{2}} n^{-3/2}, & (k+3)k^{-1} \le i \le n-2 \\ c \nu n^{-3/2}, & 0 \le i < (k+3)k^{-1}, \quad i \le n-2 \end{cases} \tag{29}$$

since when $i \le n-2$, $i < (k+3)k^{-1}$ we have

$$\tau_i \le (n/(n-i-1))^{\frac{1}{2}} \le c \, . \tag{30}$$

Note now that in the same way as we obtained (14) we can prove that (in the notations of (14))

$$\left|\int g(x) N_t * H_{(s)}(dx)\right| \le c \sup_z |g(z)| s^{-1} t \int |x| \, |Q - N|(dx)$$

(in proving this instead of (12) one uses the expansion

$$\psi(x) = \psi(0) + \sum_{\|a\|=1} \frac{x^a}{a!} D_a\psi(\theta x)) .$$

Like (14) implies (26), this implies

$$|U_{n-1} * H_1(A)| \le c \sup_x |H_{n-1}(A-x)| n^{-\frac{1}{2}} \tau_i \int |x| |P-N| (dx) . \tag{31}$$

Together with (27), (31) gives

$$|U_{n-1} * H_1(A)| \le c\bar{c}(k) n^{-1} T\nu \int |x| |P-N| (dx) . \tag{32}$$

Now

$$\int |x| |P-N| (dx) = \left(\int_{S_1} + \int_{S_1^c}\right) |x| |P-N| (dx)$$

$$\le |P-N| (R^k) + \nu \tag{33}$$

and by Lemma 1

$$|P-N| (R^k) \le ck^{-3/2} \nu^{k/(k+3)} . \tag{34}$$

Since by (23)

$$\nu^{k/(k+3)} \le c_1^{-3/(k+3)} \nu$$

and, obviously

$$c_1^{-3/(k+3)} \le c ,$$

(33), (34) imply

$$\int |x| |P-N| (dx) \le c\nu . \tag{35}$$

Together (32) and (35) give

$$|U_{n-1} * H_1(A)| \le \bar{c}(k) c\nu^2 T n^{-1} . \tag{36}$$

Summing inequalities (29) and (36), using (13) §1, we obtain

$$|(P_n - N) * N_T(A)| \le c\nu n^{-\frac{1}{2}} + \bar{c}(k) c\nu^2 T n^{-1} . \tag{37}$$

Inequality (37) together with Corollary 2 imply

$$|(P_n - N)(A)| \le c(\nu n^{-\frac{1}{2}} + \bar{c}(k)\nu^2 T n^{-1} + kT^{-1}) . \tag{38}$$

The right side in (38) attains its minimum when

$$T = \bar{c}(k)^{-\frac{1}{2}} k^{\frac{1}{2}} \nu^{-1} n^{\frac{1}{2}}$$

and with this value of T is equal to

$$\bar{c}(k)\nu n^{-\frac{1}{2}}[c(\bar{c}(k)^{-1} + \bar{c}(k)^{-\frac{1}{2}}k^{\frac{1}{2}})] . \tag{39}$$

Taking $\bar{c}(k)$ large enough to make the expression in square brackets in (39) to be ≤ 1 (which is the case when $\bar{c}(k) \sim ck$, c being an appropriate constant) we finish the proof of the first part of the theorem.

To prove the second part of the theorem again we shall use induction on n and start with the representation (cf. (24))

$$\begin{aligned} P_n - N &= P^n_{(n)} - N^n_{n^{\frac{1}{2}}} \\ &= \Big(\sum_{i=1}^{n-2} V_i + H_{n-1} + nV_0\Big) * H_1 \end{aligned} \tag{40}$$

where

$$\begin{aligned} &V_i = H_i * N_{t_i}, \quad i = \overline{0,n-2}, \quad H_i = P^i_{(n)} - N^i_{n^{\frac{1}{2}}}, \quad i = \overline{1,n-1} \\ &H_0 = \delta_0, \quad t_i = (n/(n-i-1))^{\frac{1}{2}}, \quad i = \overline{0,n-2} . \end{aligned} \tag{41}$$

Fix a Borel set B. Exactly in the same way as (29) we obtain

$$|V_i * H_1(B)| \leq \begin{cases} c_2\nu^2 t_i^3 i^{-\frac{1}{2}} n^{-3/2}, & (k+3)k^{-1} \leq i \leq n-2 \\ c\nu n^{-3/2}, & 0 \leq i < (k+3)k^{-1} . \end{cases} \tag{42}$$

Also

$$\begin{aligned} |H_{n-1} * H_1(B)| &= |\textstyle\int H_{n-1}(B-x)H_1(dx)| \\ &\leq \sup_x |H_{n-1}(B-x)| \textstyle\int |H_1|(dx) . \end{aligned}$$

By the inductive hypothesis for any Borel set E

$$\begin{aligned} |H_{n-1}(E)| &= |P^{n-1}_{(n)}(E) - N^{n-1}_{n^{\frac{1}{2}}}(E)| \\ &= |P_{n-1}((n/(n-1))^{\frac{1}{2}}E) - N((n/(n-1))^{\frac{1}{2}}E) \\ &\leq c_2\nu(n-1)^{-\frac{1}{2}} \end{aligned}$$

and by Lemma 1

$$\int |H_1|(dx) = |P - N|(R^k)$$
$$\leq ck^{-3/2}\nu^{k/(k+3)} . \qquad (43)$$

Thus

$$|H_{n-1} * H_1(B)| \leq c_2 ck^{-3/2}\nu^{(2k+3)/(k+3)} n^{-\frac{1}{2}} . \qquad (44)$$

Summing inequalities (42) and (44) and observing that

$$\sum_{i=1}^{n-2} t_i^3 \, i^{-\frac{1}{2}} \leq n^{3/2} \int_1^{n-2} (n - x - 1)^{-3/2} (x - 1)^{-\frac{1}{2}} dx + n^{3/2} (n - 2)^{-\frac{1}{2}}$$
$$\leq cn \qquad (45)$$

we obtain

$$\delta_n \leq c_2 \nu n^{-\frac{1}{2}} [c(c_2^{-1} + c_1 + c_1^{k/(k+3)} k^{-3/2})] . \qquad (46)$$

Taking c_1 small enough and c_2 large enough we can make the expression in square brackets in (46) to be less than 1. □

Theorem 2. Under the same conditions and with the same notations as in Theorem 1

1. There exist $\tilde{c}(k)$ such that for any convex Borel set A for all $n = 1,2,\ldots$

$$|P_n(A) - N(A)| \leq \tilde{c}(k)\hat{\nu} n^{-\frac{1}{2}} (1 + r^3(A))^{-1} \qquad (47)$$

where

$$r(A) = \inf\{|x|, \; x \in \partial A\} .$$

2. There exist two constants c_3, $c_3(k)$ such that if $\nu < c_3$ then for all $n = 1,2,\ldots$

$$\int |x|^3 |P_n - N|(dx) \leq c_3(k)\nu n^{-\frac{1}{2}} . \qquad (48)$$

3. There exist two constants c_4, $c_4(k)$ such that if $\nu < c_4$ then for any Borel set B for all $n = 1,2,\ldots$

$$|P_n(B) - N(B)| \leq c_4(k)\hat{\nu} n^{-\frac{1}{2}} (1 + r^3(B))^{-1} . \qquad (49)$$

The constants $\tilde{c}(k)$, $c_3(k)$, $c_4(k)$ may be taken to be

$$\tilde{c}(k) \sim \tilde{c}k^{5/2}, \quad c_3(k) \sim c_3' k^{3/2}, \quad c_4(k) \sim c_4' k^{3/2} .$$

The proof of Theorem 2 is based on the following smoothness lemma.

Lemma 4. Let P be a probability distribution on R^k with mean 0 and covariance matrix I. Denote

$$Q = P-N\,,\ R(A) = r^3(A)Q(A),\ \delta = \sup_{A\in C} |R(A)|$$

$$Q_T = (P-N)*N_T,\ R_T(A) = r^3(A)Q_T(A), \delta_T = \sup_{A\in C} |R_T(A)|.$$

Then for any $T \geq 1$

$$\delta \leq 6\delta_T + ck^{5/2}\, T^{-1} \tag{50}$$

A somewhat long proof of Lemma 4 is based on a number of auxiliary facts which we state as Lemmas 5-8.

Lemma 5. 1. For any $E \subset R^k$, $x \in R^k$, $\lambda \in R^1$

$$|r(E)-r(E+x)| \leq |x|,\ r(\lambda E) = \lambda r(E)$$

2. If $E \in R^k$, $\varepsilon > 0$ and $0 \notin E^\varepsilon$ then $r(E) - r(E^\varepsilon) = \varepsilon$

The proof of Lemma 5 is elementary and is omitted.

Lemma 6. For any probability measures P,Q on R^k, any $s > 0$ and any Borel set $B \subset R^k$

$$r^s(B)|P(B)-Q(B)| \leq \int |x|^s\, |P-Q|(dx)$$
$$\leq \int |x|^s\, (P+Q)(dx).$$

Proof The second inequality being obvious, we have to prove only the first one. Denote

$$\tilde{B} = \begin{cases} B, & \text{if } 0 \notin B \\ B^c, & \text{if } 0 \in B\ . \end{cases}$$

Then

$$r^s(B)|P(B)-Q(B)| = r^s(\tilde{B})|P(\tilde{B})-Q(\tilde{B})|$$
$$\leq r^s(\tilde{B})|P-Q|(\tilde{B})$$
$$\leq \int |x|^s\, |P-Q|(dx)\ .\ \square$$

Lemma 7. 1) For all $s,\ r = \overline{0,3},\ \tau > 0$

$$\int |x|^s |x_1|^r \phi_\tau(x)dx = c(s,r)\tau^{-s-r}\Gamma((k+s+r)/2)/\Gamma((k+r)/2)$$
$$\leq c\tau^{-s-r}k^{s/2}\ .$$

2) For P the same as in Lemma 4 and $s = \overline{0,2}$

$$\int |x|^s \, P(dx) \le k^{s/2} .$$

Proof. The first assertion of the lemma is verified by a direct calculation. The second one follows from $\int |x| P(dx) \le (\int |x|^2 P(dx))^{\frac{1}{2}} = k^{\frac{1}{2}}$. □

Lemma 8. Let $w, v = v(k)$ be such that $0 < w < 1$, $N(S_v) = w$. Let A be a set with boundary contained in the union of the boundaries of at most two convex sets. Finally let T be a number ≥ 1. Denote $t = vT^{-1}$. We also preserve the notations of Lemma 4.

(1) Suppose $|x| < t$. If (a) $0 \notin A^{2t}$ or if (b) $0 \in A$ and $r(A^t) - r(A) \le t$, then

$$r^3(A^t)Q(A^t - x) \ge R(A) - ck^{5/2}T^{-1} . \tag{51}$$

(2) For any $x \in R^k$

$$r^3(A^t)Q(A^t - x) \ge R(A^t - x) - c(k|x| + k^{\frac{1}{2}}|x|^2 + |x|^3). \tag{52}$$

Proof. (1) (a) We have

$$A \subset A^t - x \subset A^{2t} .$$

Thus

$$\begin{aligned} r^3(A^t)Q(A^t - x) &\ge r^3(A^t)(P(A) - N(A^{2t})) \\ &\ge R(A) + (r^3(A^t) - r^3(A))P(A) + r^3(A)N(A) \\ &\quad - r^3(A^t)N(A^{2t}) . \end{aligned} \tag{53}$$

Since $r(A) - r(A^t) = t$ by Lemma 5,2), we have

$$\begin{aligned} (r^3(A) - r^3(A^t))P(A) &\le 3tr^2(A)P(A) \\ &\le 3t\int |x|^2 P(dx) = 3tk . \end{aligned} \tag{54}$$

In the case we consider $r(A^{2t}) \le r(A)$, hence

$$\begin{aligned} r^3(A)N(A) - r^3(A^t)N(A^{2t}) &\ge r^3(A^{2t})N(A) - r^3(A^t)N(A^{2t}) \\ &= r^3(A^{2t})(N(A) - N(A^{2t})) + (r^3(A^{2t}) - r^3(A^t))N(A^{2t}) . \end{aligned} \tag{55}$$

By Remark 2,

$$r^3(A^{2t})(N(A) - N(A^{2t})) \le \int_{A^{2t} \setminus A} |x|^3 N(dx) \le ck^2 t . \tag{56}$$

Moreover, since $r(A^t) = r(A^{2t}) + t$,

$$(r^3(A^{2t})-r^3(A^t))N(A^{2t}) = -[3tr^2(A^{2t})+3t^2\, r(A) + t^3]N(A^{2t})$$

$$\geq -(3t\int|x|^2\, N(dx) + 3t^2 \int|x|N(dx) + t^3)$$

$$\geq -(3tk + 3t^2k^{\frac{1}{2}} + t^3). \tag{57}$$

From (53) - (57) and (19) we deduce (51).

1), (b). As in the case (a) we have (53). Now $0 \in A$ implies $r(A) \leq r(A^t)$ and and we may write

$$(r^3(A^t) - r^3(A))P(A) + r^3(A)N(A) - r^3(A^t)N(A^{2t})$$

$$= (r^3(A^t) - r^3(A))(P(A)-N(A^{2t})) + r^3(A)(N(A)-N(A^{2t}))$$

$$= (r^3(A^t) - r^3(A))(-P(A^c) + N((A^{2t})^c) - r^3(A)(N(A^{2t})-N(A))$$

$$\geq -(r^3(A^t)-r^3(A))P(A^c) - r^3(A)(N(A^{2t}) - N(A)). \tag{58}$$

Since $r(A^t) - r(A) \leq t$, by Lemma 7 we have

$$(r^3(A^t) - r^3(A))P(A^c) \leq (3tr^2(A) + 3t^2r(A) + t^3)P(A^c)$$

$$\leq 3 + \int|x|^2P(dx) + 3t^2\int|x|P(dx) + t^3$$

$$\leq 3tk + 3t^2k^{\frac{1}{2}} + t^3. \tag{59}$$

Furthermore, by Remark 2

$$r^3(A)(N(A^{2t}) - N(A)) \leq \int_{A^{2t}\setminus A} |x|^3\, N(dx) \leq ck^2t. \tag{60}$$

Relations (58) - (60) and (19) imply (51).

2) We have

$$r^3(A^t)Q(A^t-x) = R(A^t-x) + (r^3(A^t) - r^3(A^t-x))Q(A^t-x)$$

and, since by Lemma 5 $|(r(A^t) - r(A^t-x)| \leq |x|$,

$$|r^3(A^t) - r^3(A^t-x)| \leq |x|^3 + 3|x|^2\, r(A^t-x) + 3|x|r^2(A^t-x).$$

Applying now Lemma 6 and 7 we obtain

$$|r^3(A^t) - r^3(A^t-x)|Q(A^t-x) \leq c(k|x|+k^{\frac{1}{2}}|x|^2 + |x|^3),$$

which proves (52). □

<u>Proof of Lemma 4</u>. Fix $T \geq 1$, take $v = v(k)$ such that $N(x:|x| < v(k)) = 7/8$ and take $\varepsilon > 0$ such that $\varepsilon < k^{5/2}\, T^{-1}$. Denote $t = vT^{-1}$. Since $R(A) = -R(A^c)$ and

$\delta = \sup_{A \in C} |R(A)|$ we can choose A_ε such that $R(A_\varepsilon) > \delta - \varepsilon$, which is either a convex set or the complement to a convex set.

1. Suppose first that $0 \notin A_\varepsilon^{2t}$. Then

$$\delta_T \geq R_T(A_\varepsilon^t) = \left(\int_{|x|<t} + \int_{|x|\geq t}\right) r^3(A_\varepsilon^t) Q(A_\varepsilon^t - x) N_T(dx)$$

$$= I_1 + I_2.$$

By Lemma 8, 1), (a) we have

$$I_1 \geq (R(A_\varepsilon) - ck^{5/2}\, T^{-1}) N_T(|x| < t)$$

$$\geq (\delta - \varepsilon - ck^{5/2}\, T^{-1}) N(|x| < v(k))$$

$$= \frac{7}{8}(\delta - \varepsilon - ck^{5/2}\, T^{-1}).$$

By Lemma 8, 2) and Lemma 7 we may write

$$I_2 \geq \int_{|x|\geq t} (R(A_\varepsilon^t - x) - c(k|x| + k^{\frac{1}{2}}|x|^2 + |x|^3)) N_T(dx)$$

$$\geq -\delta\, N_T(|x| \geq t) - ck^{3/2}\, T^{-1}$$

$$= -\delta/8 - ck^{3/2}\, T^{-1}.$$

Thus

$$\delta_T \geq \frac{3}{4}\,\delta - ck^{5/2}\, T^{-1} - \frac{7}{8}\,\varepsilon\ . \tag{61}$$

2. Let now $0 \in A_\varepsilon^{2t}$. Since $R(A_\varepsilon) = r^3(A_\varepsilon)Q(A_\varepsilon) > \delta - \varepsilon$ and $|Q(A_\varepsilon)| \leq 1$, we have $r^3(A_\varepsilon) > \delta - \varepsilon$. We may also suppose $\delta - \varepsilon > (2t)^3$ since otherwise $\delta < \varepsilon + (2t)^3 < ck^{5/2}T^{-}$ by the choice of ε and (19), and (50) follows. Thus $r(A_\varepsilon) > 2t$. Together with $0 \in A_\varepsilon^{2t}$ this implies in particular $0 \in A_\varepsilon$. Denote $r = r(A_\varepsilon)$ and take η such that $0 < \eta < r - 2t$. We have

$$A_\varepsilon^t = (A_\varepsilon \setminus S_{r-\eta})^t \cup \bar{S}_{r-\eta-t},\quad (A_\varepsilon \setminus S_{r-\eta})^t \cap \bar{S}_{r-\eta-t} = \emptyset$$

$$r((A_\varepsilon \setminus S_{r-\eta})^t) = r-\eta-t < r.$$

Thus

$$R_T((A_\varepsilon \setminus S_{r-\eta})^t) + R_T(S_{r-\eta}^t) = (r-\eta-t)^3 (Q * N_T)(A_\varepsilon^t) - R_T(\bar{S}_{r-\eta-t}) + R(S_{r-\eta-t})$$

$$< |R_T(A_\varepsilon^t)| + |R_T(\bar{S}_{r-\eta-t})| + |R_T(S_{r-\eta+t})| < 3\delta_T. \tag{62}$$

On the other hand

$$R_T((A_\varepsilon \setminus S_{r-\eta})^t) + R_T(S^t_{r-\eta}) = (\int_{|x|<t} + \int_{|x|\geq t})[r^3((A_\varepsilon \setminus S_{r-\eta})^t)Q((A_\varepsilon \setminus S_{r-\eta})^t - x)$$

$$+ r^3(S^t_{r-\eta})Q(S^t_{r-\eta}-x)]N_T(dx) = I_3 + I_4 . \qquad (63)$$

By Lemma 8, 1), a) with $A = A_\varepsilon \setminus S_{r-\eta}$ and 1), b) with $A = S_{r-\eta+t}$

$$I_3 \geq (R(A_\varepsilon \setminus S_{r-\eta}) + R(S_{r-\eta}) - ck^{5/2}T^{-1})N_T(|x| < t)$$

$$= \frac{7}{8}((r-\eta)^3 Q(A_\varepsilon) - ck^{5/2} T^{-1})$$

$$\geq \frac{7}{8}((1 - \frac{\eta}{r})^3 (\delta-\varepsilon) - ck^{5/2} T^{-1}) . \qquad (64)$$

Note now that

$$(A_\varepsilon \setminus S_{r-\eta})^t - x = (A^t_\varepsilon \setminus \bar{S}_{r-\eta-t}) - x = (A^t_\varepsilon - x) \setminus (\bar{S}_{r-\eta-t} - x)$$

$$A^t_\varepsilon - x \supset \bar{S}_{r-\eta-t} - x$$

and

$$r((A_\varepsilon \setminus S_{r-\eta})^t - x) \leq \min (r(A^t_\varepsilon - x), r(\bar{S}_{r-\eta-t} - x)).$$

Thus

$$|R((A_\varepsilon \setminus S_{r-\eta})^t - x) + R(S^t_{r-\eta} - x)|$$

$$\leq |R(A^t_\varepsilon - x)| + |R(\bar{S}_{r-\eta-t} - x)| + |R(S^t_{r-\eta} - x)| \leq 3\delta .$$

Hence by Lemma 8, 2) and Lemma 7

$$I_4 \geq \int_{|x|\geq t} R((A_\varepsilon \setminus S_{r-\eta})^t - x) + R(S^t_{r-\eta} - x)$$

$$- c(k|x| + k^{\frac{1}{2}} |x|^2 + |x|^3)N_T(dx)$$

$$\geq -(3\delta \cdot \frac{1}{8} + ck^{3/2} T^{-1}). \qquad (65)$$

Relations (62) - (65) now imply

$$3\delta_T \geq \frac{7}{8}(1 - \frac{\eta}{r})^3 (\delta-\varepsilon) - ck^{5/2} T^{-1} - \frac{3}{8}\delta$$

and, η being arbitrary small, we obtain

$$3\delta_T \geq \frac{\delta}{2} - ck^{5/2} T^{-1} - \frac{7}{8}\varepsilon. \qquad (66)$$

Thus in all cases we have either (61) or (66), and since ε can be taken arbitrary small, this imply (50). □

In the proof of Theorem 2 we shall need yet another auxiliary fact contained in Lemma 10. To prove it we first prove

<u>Lemma 9.</u> If $Y_1,\ldots, Y_n$ are i.i.d. real random variables with $EY_i = 0$, $EY_i^2 = 1$, then

$$E\left|\sum_1^n Y_i\right|^3 \leq 4nE|Y_1|^3 + 2n^{3/2}. \tag{67}$$

<u>Proof.</u> We may suppose $E|Y_1|^3 < \infty$ and denote for brevity $\operatorname{sgn} u = f(u)$. We have

$$E\left|\sum_1^n Y_i\right|^3 = E\left(\sum_1^n Y_i\right)^2\left(\sum_1^n Y_i\right)f\left(\sum_1^n Y_i\right) = nE\left(\sum_1^n Y_i\right)^2 Y_1 f\left(\sum_1^n Y_i\right)$$

$$= nEY_1^3 f\left(\sum_1^n Y_i\right) + 2nEY_1^2\left(\sum_2^n Y_i\right)f\left(\sum_1^n Y_i\right) + nEY_1\left(\sum_2^n Y_i\right)^2 f\left(\sum_1^n Y_i\right) = I_1+I_2+I_3. \tag{68}$$

Clearly

$$|I_1| \leq nE|Y_1|^3$$

$$|I_2| \leq 2nEY_1^2 E\left|\sum_2^n Y_i\right| \leq 2n\left(E\left(\sum_2^n Y_i\right)^2\right)^{\frac{1}{2}} \leq 2n^{3/2}. \tag{69}$$

Denoting

$$E = \left\{|Y_1| \geq \left|\sum_2^n Y_i\right|\right\},\ F_1 = \left\{\sum_2^n Y_i > 0\right\},\ F_2 = \left\{\sum_2^n Y_i < 0\right\}$$

we may also write

$$I_3 = \left(\int_E + \int_{E^cF_1} + \int_{E^cF_2}\right) nY_1\left(\sum_2^n Y_i\right)^2 f\left(\sum_1^n Y_i\right)dP = J_1 + J_2 + J_3.$$

Obviously

$$|J_1| \leq nE|Y_1|^3. \tag{70}$$

Note now that on E^cF_1 we have $|Y_1| \leq \left|\sum_2^n Y_i\right| = \sum_2^n Y_i$ and hence $f\left(\sum_1^n Y_i\right) = 1$. Thus

$$J_2 = n\left(\int_{F_1} - \int_{EF_1}\right)Y_1\left(\sum_2^n Y_i\right)^2 dP .$$

Moreover denoting χ_{F_1} the indicator function of F_1 we have

$$\int_{F_1} Y_1\left(\sum_2^n Y_i\right)^2 dP = EY_1 E\left(\sum_2^n Y_i\right)^2 \chi_{F_1} = 0 .$$

Hence

$$|J_2| \leq n \int_{EF_1} |Y_1|^3\, dP \leq nE|Y_1|^3. \tag{71}$$

Exactly in the same way

$$|J_3| \leq nE|Y_1|^3 \tag{72}$$

Now (68) - (72) imply (67). □

Lemma 10. In the notations and conditions of Theorems 1-2 with $m = 0$, $V = I$, for any Borel set $B \subset R^k$ and all $n = 1,2, \ldots$

$$r^3(B)|P_n(B) - N(B)| \leq \tilde{c}(k^2 + k^{\frac{1}{2}} \nu n^{-\frac{1}{2}}).$$

Proof. By Lemma 6 with $P = P_n$, $Q = N$ and Lemma 7

$$r^3(B)|P_n(B)-N(B)| \leq \int |x|^3 (P_n+N)(dx)$$

$$\leq E|Y_n|^3 + ck^{3/2} .$$

Denoting $X_i = (X_{i1},\ldots,X_{ik}), Y_n = (Y_{n1},\ldots,Y_{nk})$ and using inequalities $(\sum_1^k \alpha_i^2)^{3/2} \leq k^{\frac{1}{2}}(\sum_1^k |\alpha_i|^3)$, $\sum_1^k \alpha_i^3 \leq (\sum_1^k \alpha_i^2)^{3/2}$ valid for any real numbers $\alpha_1,\ldots,\alpha_k$, we have by Lemma 9

$$E|Y_n|^3 \leq k^{\frac{1}{2}} E \sum_{j=1}^k |Y_{nj}|^3$$

$$\leq k^{\frac{1}{2}} \sum_{j=1}^k (4n^{-\frac{1}{2}}E|X_{ij}|^3 + 2)$$

$$\leq 4k^{\frac{1}{2}}n^{-\frac{1}{2}} E|X_1|^3 + ck^{3/2}$$

$$\leq \tilde{c}(k^2 + k^{\frac{1}{2}} n^{-\frac{1}{2}} \nu)$$

since $E|X_1|^3 = \int |x|^3(P-N+N)(dx) \leq \nu + ck^{3/2}$. □

Proof of Theorem 2. As in the proof of Theorem 1 we may suppose without loss of generality that $\mu = 0$, $V = I$, $E|X_1|^3 < \infty$.

Let us prove first (48). For this we shall use induction on n. When $n = 1$,(48) is obvious. For $n > 1$ we shall use representation (40) and notations (41).

Note that for any signed Borel measure M on R^k

$$\int |x|^3 |M|(dx) = \sup_{f \in F} | \int f(x)|x|^3 M(dx)|, \tag{73}$$

where F is the class of real Borel measurable functions with $\sup_x|f(x)| \leq 1$. Choose and fix a function f in F and denote

$$w_i(x) = \int f(x+y)|x+y|^3 V_i(dy)$$
$$= \iint f(x+y+u)|x+y+u|^3 H_i(dy)\phi_{t_i}(u)(du)$$
$$= \iint f(y+z)|y+z|^3 H_i(dy)\phi_{t_i}(z-x)dz .$$

The function w_i is a smooth function and we can define

$$\bar{w}_i(x) = w_i(x) - \sum_{\|a\|\le 2} \frac{x^a}{a!} D_a w_i(0) .$$

Since the first and the second moments of the signed measure $H_1 = P_{(n)} - N_{n^{\frac{1}{2}}}$ are equal to zero, we have

$$|\int f(x)|x|^3 V_i * H_1(dx)| = |\int w_i(x) H_1(dx)|$$
$$\le (\int_{S_1} + \int_{S_1^c})|\bar{w}_i(x)||H_1|(dx) = I_1 + I_2 . \tag{74}$$

Using now Taylor's formula (12) we may write

$$I_1 \le \int_{S_1} |\sum_{\|a\|=3} \frac{x^a}{a!} D_a w_i(\theta x)||H_1|(dx)$$
$$I_2 \le \int_{S_1^c} (|w_i(x)| + \sum_{j=0}^{2} |\sum_{\|a\|=j} \frac{x^a}{a!} D_a w_i(0)|)|H_1|(dx). \tag{75}$$

To estimate I_1, I_2 denote

$$g_i(z) = \int f(y+z)|y+z|^3 H_i(dy) .$$

Then

$$w_i(x) = \int g_i(z)\phi_{t_i}(z-x)dz$$

and a straightforward calculation shows that

$$\sum_{\|a\|=1} \frac{x^a}{a!} D_a w_i(y) = t_i^2|x|\int g_i(y+z)(x',z)\phi_{t_i}(z)dz$$
$$\sum_{\|a\|=2} \frac{x^a}{a!} D_a w_i(y) = (t_i^2|x|^2/2)\int g_i(y+z)[t_i^2(x',z)^2 - 1]\phi_{t_i}(z)dz \tag{76}$$
$$\sum_{\|a\|=3} \frac{x^a}{a!} D_a w_i(y) = (t_i^4|x|^3/6)\int g_i(y+z)(x',z)[t_i^2(x',z)^2 - 3]\phi_{t_i}(z)dz ,$$

where $x' = x/|x|$. Moreover since $(\alpha+\beta+\gamma)^3 \le 9(|\alpha|^3 + |\beta|^3 + |\gamma|^3), \alpha,\beta,\gamma \in R$

$$|g_i(y+z)| = |\int f(s+y+z)|s+y+z|^3 H_i(ds)|$$
$$\le 9(\lambda_i + (|y|^3 + |z|^3)\mu_i)$$

where $\lambda_i = \int |s|^3 |H_i|(ds), \mu_i = |H_i|(R^k)$. From the above formulas, Lemma 7 and the symmetry of ϕ_t we deduce now

$$\Big|\sum_{\|a\|=j} \frac{x^a}{a!} D_a w_i(y)\Big| \leq c|x|^j [t_i^j(\lambda_i + |y|^3 \mu_i) + t_i^{j-3} k^{3/2} \mu_i] . \tag{77}$$

Clearly

$$\begin{aligned} \int_{S_1} |x|^3 |H_1|(dx) &\leq \nu n^{-3/2} \\ \int_{S_1^c} |x|^\alpha |H_1|(dx) &\leq \int |x|^3 |H_1|(dx) \\ &= \nu n^{-3/2}, \quad 0 \leq \alpha \leq 3 \end{aligned} \tag{78}$$

and by the inductive hypothesis we have

$$\begin{aligned} \lambda_i = \int |y|^3 |H_i|(dy) &= (i/n)^{3/2} \int |y|^3 |P_i - N|(dy) \\ &\leq c_3(k) \nu i^{-\frac{1}{2}} . \end{aligned} \tag{79}$$

Relations (74) - (79) imply now for all $i = 1, \ldots, n-2$

$$I_1 + I_2 \leq c\nu n^{-3/2} [c_3(k) \nu i^{-\frac{1}{2}} t_i^3 + (k^{3/2} + t_i^3)|H_i|(R^k)] . \tag{80}$$

Note that by Theorem 1 (assuming $\nu < c_1$)

$$\mu_i = |H_i|(R^k) \leq \begin{cases} c\nu i^{-\frac{1}{2}} , & (k+3)k^{-1} \leq i \leq n-2 \\ 2 , & i < (k+3)k^{-1} , \quad i \leq n-2 . \end{cases} \tag{81}$$

Also if $i \leq n-2$, $i < (k+3)k^{-1}$ then (c.f. (33))

$$t_i = (n/(n-i-1))^{\frac{1}{2}} \leq c . \tag{82}$$

Summing now inequalities (80), taking into account (81), (82), (45) and the elementary inequality $\sum_1^{n-2} i^{-\frac{1}{2}} \leq cn^{\frac{1}{2}}$, we obtain

$$\sum_{i=1}^{n-2} \Big|\int f(x)|x|^3 V_i * H_1(dx)\Big| \leq c\nu n^{-3/2} (c_3(k)\nu n + k^{3/2} \nu n + k^{3/2}) . \tag{83}$$

By (79) and (5) we also have

$$\begin{aligned} \Big|\int f(x)|x|^3 H_{n-1} * H_1(dx)\Big| &= \Big|\iint f(x+y)|x+y|^3 H_{n-1}(dx) H_1(dy)\Big| \\ &\leq 4\int |x|^3 |H_{n-1}|(dx)|H_1|(R^k) + 4\int |y|^3 |H_1|(dy)|H_{n-1}|(R^k) \\ &\leq cc_3(k) k^{-3/2} \nu^{(2k+3)/(k+3)} n^{-\frac{1}{2}} + c\nu n^{-3/2} . \end{aligned} \tag{84}$$

Finally, (74) - (78) imply

$$n\left|\int f(x)|x|^3 V_0 * H_1(dx)\right| \le c\nu n^{-\frac{1}{2}} A(t_0,k) \tag{85}$$

where

$$A(t_0,k) = t_0^3 + k^{3/2} \sum_{j=0}^{3} t_0^{-j} .$$

Note for later use that (85) is true for any $t_0 > 0$ and any real Borel measurable f with $\sup_x |f(x)| \le 1$, and thus may be written as

$$\int |x|^3 |N_{t_0} * H_1|(dx) \le c\nu n^{-3/2} A(t_0,k) . \tag{86}$$

In the case we consider now

$$c' \le t_0 = (n/(n-1))^{\frac{1}{2}} \le c'' \quad (n \ge 2)$$

so that

$$A(t_0,k) \le ck^{3/2}. \tag{87}$$

Now (73) with $M = P_n - N$, (40) and (83) - (85) imply

$$\int |x|^3 |P_n - N|(dx) \le c_3(k)\nu n^{-\frac{1}{2}}[c(\nu + k^{-3/2}\nu^{k/(k+3)} + (1+\nu)k^{3/2}(c_3(k))^{-1}]. \tag{88}$$

Taking c_3 small enough and c_3' large enough, and assuming that $\nu \le c_3$, $c_3(k) \sim c_3' k^{3/2}$, we can make the expression in square brackets in (88) to be less than 1. This proves (48).

Estimate (49) follows easily from (48) and Theorem 1. Indeed by Lemma 6 and (48) if $\nu < c_3$ then for any Borel set B

$$r^3(B)|P_n(B) - N(B)| \le \int |x|^3 |P_n - N|(dx) \le c_3(k)\nu n^{-\frac{1}{2}} . \tag{89}$$

On the other hand by Theorem 1, if $\nu < c_1$ then

$$|P_n(B) - N(B)| \le c_2 \hat{\nu} n^{-\frac{1}{2}} . \tag{90}$$

Assuming that $\nu < c_4 = \min(c_1, c_3)$, summing (89) and (90) and noting that $\nu \le \hat{\nu}$ we obtain (49) with $c_4(k) = c_2 + c_3(k) \sim c_4' k^{3/2}$.

In view of (1) and (49) to prove (47) it is enough now to show that if

$$\nu > c_4 \tag{91}$$

then for any convex Borel set A

$$r^3(A)|P_n(A) - N(A)| \le \hat{c}(k)\nu n^{-\frac{1}{2}}, \quad n = 1,2,\dots . \tag{92}$$

Again we shall use induction on n. When $n = 1$, (92) follows immediately from Lemma 6. For $n > 1$ we shall use representation (24) with $T \ge 1$ and notations (25). Fix a convex Borel set A and denote $g_i'(z) = r^3(A)H_i(A - z)$

$$\begin{aligned} v_i(x) &= r^3(A)U_i(A - x) \\ &= \int g_i'(z)\phi_{\tau_i}(z - x)dz . \end{aligned}$$

Let

$$\tilde{v}_i(x) = v_i(x) - \sum_{\|a\| \le 2} \frac{x^a}{a!} D_a v_i(0) . \tag{93}$$

As in (74) - (75) we have

$$\begin{aligned} r^3(A)|U_i * H_1(A)| &\le \left(\int_{S_1} + \int_{S_1^c}\right)|\tilde{v}_i(x)||H_1|(dx) \\ &= I_1' + I_2' \end{aligned} \tag{94}$$

and

$$I_1' \le \int_{S_1} \Big|\sum_{\|a\|=3} \frac{x^a}{a!} D_a v_i(\theta x)\Big||H_1|(dx) \tag{95}$$

$$I_2' \le \int_{S_1^c}\Big(|v_i(x)| + \sum_{j=0}^{2} \Big|\sum_{\|a\|=j} \frac{x^a}{a!} D_a v_i(0)\Big|\Big)|H_1|(dx) . \tag{96}$$

Note that equations (76) remain true if we replace in them w_i, t_i, g_i by v_i, τ_i, g_i', respectively. Moreover, Lemma 5, 1) implies

$$\begin{aligned} |g_i'(y + z)| &\le 9(r^3(A - y - z) + |y|^3 + |z|^3)|H_i(A - y - z)| \\ &\le 9(\lambda_i' + (|y|^3 + |z|^3)\mu_i') \end{aligned}$$

where $\lambda_i' = \sup_{C} r^3(B)|H_i(B)|$, $\mu_i' = \sup_{C} |H_i(B)|$. Now in the same way as we obtained (77) we deduce

$$\Big|\sum_{\|a\|=j} \frac{x^a}{a!} D_a v_i(y)\Big| \le c|x|^j[\tau_i^j(\lambda_i' + |y|^3\mu_i') + \tau_i^{j-3}k^{3/2}\mu_i'] . \tag{97}$$

By Lemma 5, 1) and the inductive hypothesis for any convex Borel set B

$$\begin{aligned} r^3(B)|H_i(B)| &= (i/n)^{3/2}r^3((n/i)^{\frac{1}{2}}(B))|(P_i - N)((n/i)^{\frac{1}{2}} B)| \\ &\le \hat{c}(k)\nu i^{-\frac{1}{2}} , \end{aligned} \tag{98}$$

and by Theorem 1 and (91)

$$\begin{aligned} |H_i(B)| &= |(P_i - N)((n/i)^{\frac{1}{2}} B)| \\ &\le ck\nu i^{-\frac{1}{2}} . \end{aligned} \tag{99}$$

From (94) - (99), Lemma 7 and (78) we deduce (using the fact that $\tau_i \geq 1$ if $T \geq 1$)

$$r^3(A)|U_i * H_1(A)| \leq c\nu^2 n^{-3/2} i^{-\frac{1}{2}}((\hat{c}(k)+k)\tau_i^3 + k^{5/2}) . \tag{100}$$

Now (100), (13) §1 and the elementary inequality $\sum_1^n i^{-\frac{1}{2}} \leq cn^{\frac{1}{2}}$ imply

$$\sum_{i=1}^{n-2} r^3(A)|U_i * H_1(A)| \leq c[(\hat{c}(k)+k)T + k^{5/2}]\nu^2 n^{-1} . \tag{101}$$

The term $r^3(A)|U_{n-1} * H_1(A)|$ is estimated in a similar way, the difference being that instead of definition (93) we put now

$$\bar{v}_{n-1}(x) = v_{n-1}(x) - v_{n-1}(0)$$

(see also (35)). The resulting estimate is

$$r^3(A)|U_{n-1} * H_1(A)| \leq c[(\hat{c}(k)+k)T + k^{5/2}]\nu^2 n^{-1} . \tag{102}$$

Finally, by Lemma 6, (86) and the relation $2^{-\frac{1}{2}} \leq \tau_0 \leq 2^{\frac{1}{2}}$ valid when $T \geq 1$, $n \geq 2$, we have (recall that $U_0 = N_{\tau_0}$)

$$\begin{aligned} nr^3(A)|U_0 * H_1(A)| &\leq n\int |x|^3 |N_{\tau_0} * H_1|(dx) \\ &\leq ck^{3/2}\nu n^{-\frac{1}{2}} . \end{aligned} \tag{103}$$

Relation (24), (101) - (103) and Lemma 4 imply now that for any $T \geq 1$

$$\begin{aligned} r^3(A)|(P_n - N)(A)| &\leq c[(\hat{c}(k)+k)T + k^{5/2}]\nu^2 n^{-1} \\ &\quad + ck^{3/2}\nu n^{-\frac{1}{2}} + ck^{5/2}T^{-1} . \end{aligned} \tag{104}$$

The right side in (104) attains its minimum when

$$T = \frac{k^{5/4} n^{\frac{1}{2}}}{(\hat{c}(k)+k)^{\frac{1}{2}}\nu} \tag{105}$$

and with this value of T it is equal to

$$\hat{c}(k)\nu n^{-\frac{1}{2}}[c(\hat{c}(k)^{-1}(k^{5/4}(\hat{c}(k)+k)^{\frac{1}{2}} + k^{5/2}\nu n^{-\frac{1}{2}} + k^{3/2})] . \tag{106}$$

If T defined by (105) is ≥ 1, i.e.

$$\nu n^{-\frac{1}{2}} \leq k^{5/4}(\hat{c}(k)+k)^{-\frac{1}{2}}$$

then (106) is majorized by

$$\hat{c}(k)\nu n^{-\frac{1}{2}}[c\hat{c}(k)^{-1}(k^{5/4}(\hat{c}(k)+k^{5/2})\hat{c}(k)^{-\frac{1}{2}} + k^{3/2})] . \tag{107}$$

On the other hand by Lemma 10, $r^3(A)|(P_n - N)(A)|$ is always not greater than

$$\tilde{c}(k^2 + k^{\frac{1}{2}}\nu n^{-\frac{1}{2}}) \ . \tag{108}$$

Thus if T defined by (105) is ≤ 1, i.e.

$$1 \leq k^{-5/4}(\hat{c}(k) + k)^{\frac{1}{2}}\nu n^{-\frac{1}{2}} \ ,$$

then (108) is majorized by

$$\hat{c}(k)\nu n^{-\frac{1}{2}}[\tilde{c}\hat{c}(k)^{-1}(k^{3/4}(\hat{c}(k) + k)^{\frac{1}{2}} + k^{\frac{1}{2}})] \ . \tag{109}$$

Assuming that $\hat{c}(k)$ has been chosen such that the expressions in square brackets in (107) and (109) are ≤ 1 (this is the case when $\hat{c}(k) \sim \hat{c}k^{5/2}$, where $\hat{c}$ is an appropriate constant), we obtain (92) whatever be (≥ 1 or ≤ 1) T defined by (105). □

Remark 3. The moment conditions in Theorems 1 and 2 can be weakened considerably. To indicate a result in this direction, preserving the notations of Theorems 1 and 2, define

$$\nu_{1n} = \int_{|x| \leq n^{\frac{1}{2}}} |x|^3 |\bar{P} - N|(dx) + n^{\frac{1}{2}} \int_{|x| > n^{\frac{1}{2}}} |x|^2 |\bar{P} - N|(dx)$$

and for a Borel set $B \subset R^k$ let

$$\nu_{2n} = \int_{|x| \leq n^{\frac{1}{2}} r(B)} |x|^3 |\bar{P} - N|(dx) + (n^{\frac{1}{2}} r(B))^3 \int_{|x| > n^{\frac{1}{2}} r(B)} |\bar{P} - N|(dx) \ .$$

Define also the class Γ of non-negative functions on $[0,\infty)$ by $g \in \Gamma$ iff $g(0) = 0, g(1) = 1$, $g(x)$ and $x/g(x)$ are non-decreasing, and for a $g \in \Gamma$ put

$$\lambda_g = \int |x|^2 g(|x|)|\bar{P} - N|(dx) \ .$$

Then for $n \geq 5 - \min(3,k)$

1. There exist $c'(k)$ such that for any convex Borel set A

$$|P_n(A) - N(A)| \leq c'(k)(\nu_{1n} + \nu_{2n})n^{-\frac{1}{2}}(1 + r^3(A))^{-1} \ .$$

2. There exist two constants c_5, $c_5(k)$ such that if $\nu_{1n} < c_5$ then for any $g \in \Gamma$

$$\int (1+|x|^2 g_n(|x|))|P_n - N|(dx) \le c_5(k)[\nu_{1n} n^{-\frac{1}{2}} + \lambda_g (g(n^{-\frac{1}{2}}))^{-1}],$$

where $g_n(\alpha) = g(\alpha n^{\frac{1}{2}})/g(n^{\frac{1}{2}})$.

3. There exist two constants c_6, $c_6(k)$ such that if $\nu_{1n} < c_6$ then for any Borel set B

$$|P_n(B) - N(B)| \le c_6(k)(\nu_{1n} + \nu_{2n})n^{-\frac{1}{2}}(1 + r^3(B))^{-1}.$$

Here we may take $c'(k) \sim c' k^{5/2}$, $c_5(k) \sim c_5' k^{3/2}$, $c_6(k) \sim c_6' k^{3/2}$.

§ 3. The method of characteristic functions. Integral type estimates.

The method of characteristic functions in application to the estimation of the speed of convergence in the central limit theorem has already been illustrated in § 1. Now we shall apply this method to obtain an estimate in R^k of the type described in § 1 as direction 4 of generalizations of the Berry-Esseen theorem.

Theorem 1. Let $X_1, X_2, \ldots$ be a sequence of independent random variables with values in R^k with the same distribution P. Suppose that the covariance matrix of P is I(this assumption is made only for simplicity of formulas). Denote P_n the distribution of the normalized sum $n^{-\frac{1}{2}} \sum_1^n (X_j - \mu)$, where $\mu = EX_1$, and let N be the standard normal distribution on R^k. Then there exist $c(k)$ such that for any real bounded Borel measurable function f

$$\left|\int f(x)(P_n - N)(dx)\right| \le c(k)[\omega_f(R^k)\beta_3 n^{\frac{1}{2}} + \omega_f^*(\varepsilon_n, N)], \tag{1}$$

where $\omega_f(R^k)$, $\omega_f^*(\varepsilon_n, N)$ are defined by (20), (21)*, $\beta_3 = E|X_1 - \mu|^3$ and $\varepsilon_n = c'(k)\beta_3 n^{-\frac{1}{2}}$

Remark 1. If $f(x) = \chi_A(x)$, where A is a Borel set then

$$\omega_f(x,\varepsilon) = \begin{cases} 1 & \text{if} \quad x \in (\partial A)^\varepsilon \\ 0 & \text{if} \quad x \notin (\partial A)^\varepsilon \end{cases}$$

and

$$\bar{\omega}_f(\varepsilon, N) = N((\partial A)^\varepsilon)$$

$$\omega_f^*(\varepsilon, N) = \sup\{N((\partial(A+y))^\varepsilon), y \in R^k\}$$

If moreover A is convex, then by Lemma 2, § 2

$$\omega_f^*(\varepsilon, N) \le c(k)\varepsilon, \varepsilon > 0 \ ,$$

and the right side in (1) reduces to $c(k)\beta_3 n^{-\frac{1}{2}}$. Thus (1) implies the multidimensional version of the Berry-Esseen theorem (see p.10).

* of Section 2

The proof of Theorem 1 will be based on several lemmas which we shall prove first. Without loss of generality in what follows we assume $\mu = 0$.

Lemma 1. Let $r > 0$ and $\alpha \in (0,1)$. There exist a probability measure K on R^k such that

1) $K(S_1(0)) > \alpha$

2) $\int |x|^r K(dx) < \infty$

3) $\hat{K}(t) = \int e^{i(t,x)} K(dx) = 0$ if $|t| > \beta = \beta(\alpha,r,k)$.

Proof. If s is a positive even number, then the probability distribution $D_{(s)}$ on R' with density

$$d_{(s)}(x) = c(s) \left(\frac{\sin x}{x} \right)^s ,$$

where

$$c(s) = \left(\int \left(\frac{\sin x}{x} \right)^s dx \right)^{-1} ,$$

has the characteristic function $\hat{D}_{(s)}$ which vanishes outside the interval $(-s,s)$. To show this one may observe that $(x^{-1} \sin x)^s$ is the Fourier transform of the s-fold convolution of the uniform distribution on $[-s,s]$ with itself and then consider the inverse Fourier transform.

Denote by q the smallest even integer greater than r and let $Z_1,\dots,Z_k$ be i.i.d. real random variables each with distribution $D_{(q+2)}$. Put $Z = (Z_1,\dots,Z_k)$.

Obviously one can find a positive γ such that $P(|Z| < \gamma) > \alpha$. Take K to be the distribution of $\gamma^{-1}Z$. Then we have

$$\begin{aligned} K(S_1(0)) &= P(|\gamma^{-1}Z| < 1) \\ &> \alpha \end{aligned}$$

and

$$\begin{aligned} \int |x|^q K(dx) &= E|\gamma^{-1}Z|^q \\ &\leq \gamma^{-q} k^{q/2-1} E \sum_1^k |Z_j|^q \\ &= \gamma^{-q} k^{q/2} E|Z_1|^q \\ &= \gamma^{-q} k^{q/2} c(q) \int x^{-2} (\sin x)^{q+2} dx \\ &< \infty . \end{aligned}$$

Finally

$$\hat{K}(t) = Ee^{i(t,\gamma^{-1}Z)}$$
$$= \prod_{j=1}^{k} Ee^{i(\gamma^{-1}t_j,Z_j)}$$

and

$$Ee^{i(\gamma^{-1}t_j,Z_j)} = \hat{D}_{(q+2)}(\gamma^{-1}t_j)$$
$$= 0 \quad \text{if} \quad |t_j| > \gamma(q+2)$$

which implies $\hat{K}(t) = 0$ if t is outside the cube $\{t \quad R^k: |t_j| \leq \gamma(q+2), j=\overline{1,k}\}$ and, consequently, if $|t| > k^{\frac{1}{2}}\gamma(q+2)$. □

Lemma 2. Let f be a real valued function in $L^1(R^k)$ satisfying

$$\int |x|^k \, |f(x)| dx < \infty \quad .$$

Then

$$|f|_1 \leq c(k) \max_{||a||=0,k} \int |D_a\hat{f}(t)| dt \tag{2}$$

where $\hat{f}$ is the Fourier transform of f,

$$\hat{f}(t) = \int e^{i(t,x)} f(x) dx \; .$$

Proof. Let $E = \{x: f(x) \geq 0\}$ and for a vector $b = (b_1, \ldots, b_k)$ with $b_j = 0$ or 1 put

$$Q_b = \{x = (x_1,\ldots,x_k) \in R^k, \; x_j \geq 0 \quad \text{if} \quad b_j = 0,$$
$$x_j < 0 \quad \text{if} \quad b_j = 1\}.$$

We have, denoting $||x|| = \sum_1^k |x_j|$ for $x \in R^k$

$$|f|_1 = \int |f(x)| dx$$
$$= \sum_b \Big(\int_{E \cap Q_b} - \int_{E^c \cap Q_b} \Big)(1 + (\sum_1^k (-1)^{b_j} x_j)^{k+1}) f(x) \frac{dx}{1+||x||^{k+1}} . \tag{3}$$

Further, we may suppose $D_a\hat{f}(t) \in L'$ for a such that $||a|| = 0$ or $k+1$ since otherwise the lemma is obvious. In this case by the Fourier inversion formula

$$x_1^{a_1} \ldots x_k^{a_k} f(x) = (2\pi)^{-k} \int e^{-i(t,x)} D_a \hat{f}(t) dt$$

and we may write

$$|1 + (\sum_1^k (-1)^{b_j} x_j)^{k+1} f(x)| \leq c'(k) \max_{||a||=0,k+1} \int |D_a\hat{f}(t)| dt. \tag{4}$$

Now (3) and (4) imply (2) since

$$\int \frac{dx}{1+||x||^{k+1}} \leq \int \frac{dx}{1+|x|^{k+1}} < \infty . \square$$

Let now $X_1, X_2, \ldots, X_n$ be as in Theorem 1 and let

$$Y_j = \begin{cases} X_j & \text{if} \quad |X_j| \leq n^{\frac{1}{2}} \\ 0 & \text{if} \quad |X_j| > n^{\frac{1}{2}} \end{cases} \qquad j = \overline{1,n}$$

The random variables Y_j are often called the truncated random variables. Denote also $Z_j = Y_j - EY_j$ and let W be the covariance matrix of Z_1.

Lemma 3. Denote $\tilde{\Phi}$ the density of normal $(0,W)$ distribution on R^k and let $\alpha_n = n^{\frac{1}{2}} EY_1$. If

$$\beta_3 = |X_1|^3 \leq (8k)^{-1} n^{\frac{1}{2}} \tag{5}$$

then W is nonsingular and

$$|\phi(x) - \tilde{\phi}(x)| \leq c(k)\beta_3 n^{-\frac{1}{2}}(1+|x|^2)e^{-|x|^2/3} \tag{6}$$

$$|\phi(x+\alpha_n) - \phi(x)| \leq c(k)\beta_3 n^{-\frac{1}{2}}(1+|x|)e^{-|x|^2/2+|x|/8k^{\frac{1}{2}}} \tag{7}$$

Proof. Denoting w_{ij} (resp. δ_{ij}) the elements of W(resp. of I) we have

$$\begin{aligned} |w_{ij} - \delta_{ij}| &= |EZ_{1i} Z_{1j} - EX_{1i} X_{1j}| \\ &= |EY_{1i} Y_{1j} - EY_{1i} EY_{1j} - EX_{1i} X_{1j}| \\ &\leq |E(Y_{1i} Y_{1j} - X_{1i} X_{1j})| + |EY_{1i} EY_{1j}| . \end{aligned}$$

From the definition of truncated variables Y_i it follows that

$$\begin{aligned} |E(Y_{1i} Y_{ij} - X_{1i} X_{1j})| &= \Big| \int_{|X_1| > n^{\frac{1}{2}}} X_{1i} X_{1j} \, dP \Big| \\ &\leq \int_{|X_1| > n^{\frac{1}{2}}} |X_1|^2 \, dP \\ &\leq n^{-\frac{1}{2}} \beta_3 \end{aligned}$$

and

$$|EY_{1i}| = |E(Y_{1i}-X_{1i})|$$

$$= |\int_{|X_1|>n^{\frac{1}{2}}} X_{1i}\, dP|$$

$$\leq (\int_{|X_1|>n^{\frac{1}{2}}} |X_1|^2 dP)^{\frac{1}{2}} \tag{8}$$

so that

$$|EY_{1i}EY_{1j}| \leq \int_{|X_1|>n^{\frac{1}{2}}} |X_1|^2 dP$$

$$\leq n^{-\frac{1}{2}}\, \beta_3 \quad .$$

Thus

$$|w_{ij} - \delta_{ij}| \leq 2\beta_3\, n^{-\frac{1}{2}}. \tag{9}$$

Now for any $y = (y_1,\ldots,y_k) \in R^k$

$$|((W-I)y,y) = |\sum_{i,j=1}^{k} (w_{ij}-\delta_{ij})y_i y_j|$$

$$\leq 2\beta_3 n^{-\frac{1}{2}}(\sum_1^k |y_j|)^2$$

$$\leq 2k\beta_3 n^{-\frac{1}{2}}|y|^2 \;,$$

so that when (5) is satisfied

$$|((W-I)y,y)| \leq (1/4)\; |y|^2 \tag{10}$$

and

$$(3/4)|y|^2 \leq (Wy,y) = |y|^2 + ((W-I)y,y) \leq (5/4)|y|^2.$$

Thus when (5) is satisfied

$$||W-I|| = \sup_{|y|=1} |((W-I)y,y)|$$

$$\leq 2k\beta_3 n^{-\frac{1}{2}}$$

$$\leq 1/4 \tag{11}$$

$$3/4 \leq \inf_{|y|=1} (Wy,y) \leq ||W|| = \sup_{|y|=1} (Wy,y) \leq 5/4$$

Consequently W is nondegenerate and

$$4/5 \leq ||W^{-1}|| = (\inf_{|y|=1} (Wy,y))^{-1} \leq 4/3 \tag{12}$$

$$\begin{aligned}||W^{-1}-I|| &= ||(W-I)W^{-1}|| \\ &\le ||W-I||\ \ ||W^{-1}|| \\ &\le (8/3)k\beta_3 n^{-\frac{1}{2}} \\ &\le 1/3.\end{aligned}$$

Using elementary inequality

$$|1-\alpha e^{\beta}| \le (|1-\alpha|+|\beta|)e^{|\beta|},\quad \alpha,\beta \in R^1,$$

we may write

$$\begin{aligned}|\phi(x)-\tilde{\phi}(x)| &\le (2\pi)^{-k/2}e^{-|x|^2/2}|1-|W|^{-\frac{1}{2}}\exp\{(1/2)(|x|^2-(W^{-1}x,x))\}| \\ &\le (2\pi)^{-k/2}e^{-|x|^2/2}(|1-|W|^{-\frac{1}{2}}|+|w|)e^{|w|} \qquad (13)\end{aligned}$$

where

$$\begin{aligned}w &= (1/2)(|x|^2-(W^{-1}x,x)) \\ &= ((I-W^{-1})x,x)/2\ .\end{aligned}$$

For w by (12) we have

$$\begin{aligned}|w| &\le ||I-W^{-1}||\ |x|^2/2 \\ &\le (4/3)k\beta_3\ n^{-\frac{1}{2}}|x|^2 \\ &\le |x|^2/6. \qquad (14)\end{aligned}$$

On the other hand, obviously

$$|1-|W|^{-\frac{1}{2}}| \le |W|^{-\frac{1}{2}}|1-|W||.$$

Now by (12)

$$|W| \ge (\inf_{|y|=1}(Wy,y))^k \ge (3/4)^k\ .$$

Also

$$\begin{aligned}|1-|W|\ | &= |1-|I-W+I|| \\ &\le c(k)\ \beta_3 n^{-\frac{1}{2}}\end{aligned}$$

by (9) and (5). Thus

$$|\ 1-|W|^{-\frac{1}{2}}| \le c(k)\beta_3 n^{-\frac{1}{2}}. \qquad (15)$$

From (13) - (15) we deduce now

$$\begin{aligned}|\phi(x)-\tilde{\phi}(x)| &\le c(k)e^{-|x|^2/2}(\beta_3 n^{-\frac{1}{2}}+\beta_3 n^{-\frac{1}{2}}|x|^2)e^{|x|^2/6} \\ &\le c(k)\beta_3 n^{-\frac{1}{2}}(1+|x|^2)e^{-|x|^2/3}\ ,\end{aligned}$$

i.e. (6) is satisfied.

To prove (7) we note that as in (8)

$$|EY_{1i}| \leq \int_{|X_1|>n^{\frac{1}{2}}} |X_1| dP$$

$$\leq n^{-1}\beta_3 ,$$

so that taking into consideration (5) we have

$$|\alpha_n| = n^{\frac{1}{2}}|EY_1|$$

$$\leq k^{\frac{1}{2}}\beta_3 n^{-\frac{1}{2}}$$

$$\leq 8^{-1} k^{-\frac{1}{2}} . \tag{16}$$

On the other hand

$$|\phi(x+\alpha_n)-\phi(x)| = \phi(x)|\exp\{(1/2)(|x|^2-|x+\alpha_n|^2\}-1|$$

$$\leq \phi(x)(1/2)||x|^2-|x+\alpha_n|^2|\exp\{(1/2)||x|^2-|x+\alpha_n|^2|\} \tag{17}$$

and by (16)

$$||x|^2-|x+\alpha_n|^2| = |\sum_{i=1}^{k} (2\alpha_{ni}x_i + \alpha_{ni}^2)|$$

$$\leq 2|\alpha_n| |x| + |\alpha_n|^2$$

$$\leq 2|x|k^{\frac{1}{2}}\beta_3 n^{-\frac{1}{2}} + k^{\frac{1}{2}}\beta_3 n^{-\frac{1}{2}} 8^{-1}k^{-\frac{1}{2}}$$

$$= (2k^{\frac{1}{2}}|x| + 8^{-1})\beta_3 n^{-\frac{1}{2}}$$

$$\leq 4^{-1} k^{-\frac{1}{2}}|x| + (64k)^{-1} . \tag{18}$$

Inequalities (17) - (18) imply (7). □

<u>Lemma 4.</u> We preserve the notations of Lemma 3 and denote g the characteristic function of Z_1. If

$$|t| < n^{\frac{1}{2}}/16\beta_3$$

then

$$|D_a g^n(tn^{-\frac{1}{2}})| \leq c(k)(1+|t|^{||a||})\exp\{-\frac{1}{2}\{(Wt,t) + \frac{|t|^2}{6}\}$$

for all nonnegative integral vectors a satisfying $||a|| \leq k+1 < n$.

<u>Proof.</u> First consider the case a = 0. Note that if U is a real random variable with mean $0, E|U|^3 < \infty$ and $\bar{U}$ is distributed as U and is independent of U, then

$$E|U-\bar{U}|^3 \leq 4E|U|^3 . \tag{19}$$

(this is proved easily by taking expectations in the obvious inequality $|U-\bar{U}|^3 \leq (U-\bar{U})^2 (|U| + |\bar{U}|)$).

Let now $\bar{Z}_1$ be independent of Z_1 and with the same distribution as Z_1, and put $U = (t, Z_1-\bar{Z}_1)$. We have by (19) of §1, (19) and the elementary inequality $1+\alpha \leq e^{\alpha}, \alpha$ R

$$
\begin{aligned}
|g(tn^{-\frac{1}{2}})|^2 &= E \exp\{i(tn^{-\frac{1}{2}}, Z_1-\bar{Z}_1)\} \\
&= 1-n^{-1}E(t,Z_1)^2 + \theta 6^{-1} n^{-3/2} E|(t,Z_1-\bar{Z}_1)|^3 \\
&\leq 1-n^{-1}E(t,Z_1)^2 + (2/3)n^{-3/2} E|(t,Z_1)|^3 \\
&\leq e^A,
\end{aligned}
$$

where

$$A = -n^{-1} E(t,Z_1)^2 + (2/3)n^{-3/2} E|(t,Z_1)|^3.$$

Thus for $s = 1,\ldots,n$

$$
\begin{aligned}
|g(tn^{-\frac{1}{2}})|^{2s} &\leq \exp\{sA\} \\
&\leq \exp\{nA\}\exp\{(s-n)A\} \\
&= I_1 I_2.
\end{aligned}
$$

Since

$$
\begin{aligned}
\exp\{-A\} &\leq \exp\{n^{-1}(E|(t,Z_1)|^3)^{2/3} - (2/3)n^{-3/2}E|(t,Z_1)|^3\} \\
&\leq \sup_{\alpha \geq 0} \exp\{\alpha^2 - (2/3)\alpha^3\} \\
&= e^{1/3},
\end{aligned}
$$

we have

$$I_2 \leq \exp\{(n-s)/3\}.$$

To estimate I_1 we note that for any $q \geq 1$

$$
\begin{aligned}
E|Z_1|^q &= E|Y_1 - EY_1|^q \\
&\leq 2^{q-1} E(|Y_1|^q + |EY_1|^q) \\
&\leq 2^q E|Y_1|^q \\
&\leq 2^q E|X_1|^q, \qquad (20)
\end{aligned}
$$

hence

$$(2/3)n^{-\frac{1}{2}}E|(t,Z_1)|^3 \le (2/3)|t|^3 n^{-\frac{1}{2}} E|Z_1|^3$$
$$\le (16/3)\ |t|^3 n^{-\frac{1}{2}} \beta_3$$
$$\le |t|^2/3 \quad \text{if} \quad |t| < n^{\frac{1}{2}}/16\beta_3,$$

and we obtain

$$I_1 \le \exp\{-(Wt,t) + |t|^2/3\}.$$

Thus we have for $s = 1,\ldots,n$

$$|g(tn^{-\frac{1}{2}})|^{2s} \le \exp\{-(Wt,t) + |t|^2/3\} \exp\{(n-s)/3\}, \tag{21}$$

which implies the lemma for $a = 0$.

Note now that by (20) we may write for any integer $m, 1 \le m \le k$

$$\left|\frac{\partial}{\partial t_m} g(tn^{-\frac{1}{2}})\right| = n^{-\frac{1}{2}}|EZ_{1m}\exp\{i(tn^{-\frac{1}{2}},Z_1)\}|$$
$$= n^{-\frac{1}{2}}|EZ_{1m}[\exp\{i(t\,n^{-\frac{1}{2}},Z_1)\}-1]|$$
$$\le n^{-\frac{1}{2}}\, E|Z_{1m}||(tn^{-\frac{1}{2}},Z_1)|$$
$$\le n^{-1}\ |t|E|Z_1|^2$$
$$\le 4n^{-1}\ |t|\ E|X_1|^2$$
$$= 4kn^{-1}\ |t|\ . \tag{22}$$

Furthermore, since for any integer $q \ge 2$

$$E|Y_1|^q \le n^{(q-2)/2}\, E\,|Y_1|^2$$
$$\le kn^{(q-2)/2}$$

we have, by (20), for a nonnegative integral vector b with $||b|| \ge 2$

$$|D_b g(tn^{-\frac{1}{2}})| = |D_b\, E\, \exp\{i(tn^{-\frac{1}{2}},Z_1)\}$$
$$\le n^{-||b||/2}\, E\, |Z_1^b|$$
$$\le n^{-||b||/2}\, E|\ Z_1\ |^{||b||}$$
$$\le n^{-||b||/2}\, 2^{||b||}E|Y_1|^{||b||}$$
$$\le kn^{-1}\, 2^{||b||}\ .$$

Thus we always have for any nonnegative integral vector b with $||b|| \ge 1$

$$|D_b g(tn^{-\frac{1}{2}})| \le c(k)n^{-1}\ \overline{|t|} \tag{23}$$

where $c(k) = k2^{k+1}$, $\overline{|t|} = \max(1,|t|)$.

Let us note now that $D_a|g(tn^{-\frac{1}{2}})|$ is bounded from above by a sum of terms (and the number of the terms is bounded by a $c(k)$)

$$c(k) \frac{n!}{\gamma_0!} |g^{\gamma_0}(tn^{-\frac{1}{2}}) D_{b_1} g(tn^{-\frac{1}{2}}))^{\gamma_1} \ldots (D_{b_r} g(tn^{-\frac{1}{2}}))^{\gamma_r}|$$

where $r, \gamma_0, \gamma_1, \ldots, \gamma_r$ are positive integers, $b_1, \ldots, b_r$, $b_1 \neq \ldots \neq b_r$, are non-negative integral vectors in R^k with $\|b_j\| \geq 1$ and

$$\sum_0^r \gamma_j = n, \sum_1^r \gamma_j b_j = a. \tag{24}$$

By (21) and (23) such a term is not greater than

$$c(k)(n!/\gamma_0!)\exp\{-(1/2)(Wt,t) + (1/6)|t|^2\}\exp\{(n-\gamma_0)/6\}(n^{-1}\overline{|t|})^{\sum_1^r \gamma_j}.$$

Since (24) imply

$$(n!/\gamma_0!)n^{-\sum_1^r \gamma_j} \leq 1, \sum_1^r \gamma_j = n - \gamma_0 \leq \|a\| \leq k+1 ,$$

the assertion of the lemma follows. □

Let us introduce now some new notations. Let X be a random variable in R^k with $E|X|^m < \infty$. Denoting f the characteristic function of X we have

$$f(t) = 1 + \sum_{1 \leq \|a\| \leq m} \mu_a \frac{(it)^a}{a!} + O(|t|^m)$$

$$\log f(t) = \sum_{1 \leq \|a\| \leq m} \chi_a \frac{(it)^a}{a!} + O(|t|^m) \tag{25}$$

as $t \to 0$, μ_a and χ_a being the moments and the cumulants respectively of order $a = (a_1, \ldots, a_k)$ of X. Using the relation

$$\log f(t) = \sum_{j=1}^{\infty} (-1)^{j+1} \frac{(f(t)-1)^j}{j!} \tag{26}$$

valid for small t, we can express the cumulants in terms of moments. In order to do this it is convenient to use the formal relation

$$\sum_{\|a\| \geq 1} \chi_a \frac{(it)^a}{a!} = \sum_{j=1}^{\infty} (-1)^{j+1} \frac{1}{j!} \Big(\sum_{\|a\| \geq 1} \mu_a \frac{(it)^a}{a!} \Big)^j \tag{27}$$

(cf. (25) and (26)) which may be written when the moments and cumulants of all orders exist. Equating coefficients at the same powers of t we have

$$\chi_a = \sum_{r=1}^{||a||} \sum c(a_1,\ldots,a_r;j_1,\ldots,j_r)\mu_{a_1}^{j_1} \cdots \mu_{a_r}^{j_r} \tag{28}$$

where the second summation is over all r-tuples of nonnegative integral vectors $(a_1,\ldots,a_r)$ and r-tuples of nonegative integers $(j_1,\ldots,j_r)$ satisfying

$$\sum_{s=1}^{r} j_s\, a_s = a.$$

Note that (28) implies easily that if $EX = 0$ then

$$\chi_a = \mu_a \quad \text{for all } a \text{ with } ||a|| = 1,2. \tag{29}$$

Define polynomials $\chi_r(z)$, $r = 1,2,\ldots$, $z = (z_1,\ldots,z_k)$ in k variables $z_1,\ldots,z_k \in C$ by

$$\chi_r(z) = r! \sum_{||a||=r} \frac{\chi_a}{a!} z^a . \tag{30}$$

In terms of these polynomials we can write

$$\log f(t) = 1 + \sum_{r=1}^{m} \frac{\chi_r(it)}{r!} + o(|t|^m) .$$

Define now polynomials $P_r(z)$, $r=1,2,\ldots,z \in C^k$ by the formal relation

$$1 + \sum_{r=1}^{\infty} P_r(z)u^r = \exp \{ \sum_{s=1}^{\infty} \frac{\chi_{s+2}(z)}{(s+2)!} u^s\}$$

$$= 1 + \sum_{v=1}^{\infty} \frac{1}{v!} (\sum_{s=1}^{\infty} \frac{\chi_{s+2}(z)}{(s+2)!} u^s)^v .$$

Equating coefficients at the same powers of u we have

$$P_o(z) \equiv 1$$

$$P_r(z) = \sum_{v=1}^{r} \frac{1}{v!} \sum_{j_1,\ldots,j_v}' \frac{\chi_{j_1+2}(z)}{(j_1+2)!} \cdots \frac{\chi_{j_v+2}(z)}{(j_v+2)!}$$

$$= \sum_{v=1}^{r} \frac{1}{v!} [\sum_{j_1,\ldots,j_v}' (\sum'' \frac{\chi_{d_1}\cdots\chi_{d_v}}{d_1!\ldots d_v!} z^{d_1+\ldots+d_v})] \tag{31}$$

where the summation Σ' is over all v-tuples of positive integers $(j_1,\ldots,j_v)$ satisfying

$$\sum_{i=1}^{v} j_i = r, \; j_i = 1,\ldots,r, \; 1 \le i \le v \tag{32}$$

and the summation Σ'' is over all v-tuples of nonnegative integral k-dimensional vectors $(d_1,\ldots,d_v)$ satisfying

$$||d_i|| = j_i + 2, \ 1 \leq i \leq v. \tag{33}$$

From equation (31) it follows immediately that the orders of terms of the polynomials $P_r(z)$ vary between r+2 and 3r and the coefficients of $P_r(z)$ only involve χ_d's with $||d|| \leq r+2$.

The polynomials $P_r(z)$ arise naturally in the following way. Let f be the characteristic function of a random variable X with values in R^k, with EX = 0, covariance matrix V and $E|X|^m < \infty$, $m \geq 3$. Then for the characteristic function $f^n(tn^{-\frac{1}{2}})$ of the normalized sum $n^{-\frac{1}{2}} \sum_1^n X_j$, where X_j are i.i.d. with the same distribution as X, we have (see (25), (29), (30))

$$\log f^n(tn^{-\frac{1}{2}}) = n \log f(tn^{-\frac{1}{2}})$$

$$= -\frac{1}{2}(Vt,t) + \sum_{s=1}^{m-2} \frac{\chi_{s+2}}{(s+2)!} n^{-s/2} + o(|tn^{-\frac{1}{2}}|^m)$$

as $tn^{-\frac{1}{2}} \to 0$. It follows that for any fixed $t \in R^k$

$$f^n(tn^{-\frac{1}{2}}) = \exp\{-\frac{1}{2}(Vt,t)\}\exp\{\sum_{s=1}^{m-2} \frac{\chi_{s+2}(it)}{(s+2)!} n^{-s/2} + o(n^{-(m-2)/2})\}$$

$$= \exp\{-\frac{1}{2}(Vt,t)\}(1 + \sum_{r=1}^{m-2} n^{-r/2} P_r(it))(1+o(n^{-(m-2)/2}))$$

as $n \to \infty$.

Now we can formulate our next lemma.

<u>Lemma 5</u>. Let X be a random variable with values in R^k such that $EX = 0, E|X|^m < \infty$ for some integer $m \geq 3$, and let P_r be the polynomials defined by (31) with cumulants of X. For any nonnegative integer $r \leq m-2$ and any nonnegative integral vector $a = (a_1,\ldots,a_k)$ with $||a|| \leq 3r$ we have

$$|D_a P_r(it)| \leq c(r)(1 + \beta_2^{r(m-3)/(m-2)})(1+|t|^{3r-||a||})\beta_m^{r/(m-2)},$$

where $\beta_s = E|X|^s$. If $||a|| > 3r$ then $D_a P_r(it) \equiv 0$.

<u>Proof</u>. Replacing in (31) z by it and differentiating we obtain

$$D_a P_r(it) = \sum_{v=1}^{r} \frac{i^{r+2v}}{v!} [\sum_{j_1,\ldots,j_v}{}' (\sum{}'' \frac{\chi_{d_1}\cdots\chi_{d_v}}{d_1!\ldots d_v!} \times$$

$$\times \frac{(d_1+\ldots+d_v)!}{(d_1+\ldots+d_v-a)!} t_1^{d_1+\ldots+d_v-a})] \tag{34}$$

(terms for which a component of the vector $d_1+\dots+d_v - a$ is negative are zero).

To estimate the right side of (34) let us look at the formula (28) relating cumulants and moments. For any nonnegative integral vector $d_u = (d_{u1},\dots,d_{uk})$ we have

$$\begin{aligned} |\mu_{d_u}| &= |EX_1^{d_{u1}} \dots X_k^{d_{uk}}| \\ &\le E|X|^{||d_u||} \\ &= \beta_{||d_u||} \end{aligned}$$

(here $X_1,\dots,X_k$ are the components of X). Thus, using notations of (28), by well known moment inequalities we may write

$$\begin{aligned} |\mu_{d_1}^{j_1} \dots \mu_{d_r}^{j_r}| &\le \beta_{||d_1||}^{j_1} \dots \beta_{||d_r||}^{j_r} \\ &\le \beta_{||d||}^{j_1||d_1||/||d||} \cdots\cdots\cdot \beta_{||d||}^{j_r||d_r||/||d||} \\ &= \beta_{||d||} , \end{aligned}$$

and it follows that

$$|\chi_d| \le c(d)\beta_{||d||} . \tag{35}$$

Let us prove now the following statement: the function

$$r \to (\beta_r/\beta_2^{r/2})^{1/(r-2)} \text{ is nondecreasing on } (2,m]. \tag{36}$$

It is enough to prove (36) when $\beta_2 = 1$ since the general case is reduced to this one by replacing X by $X/\beta_2^{1/2}$. The function $r \to \beta_r^{1/(r-2)}$ is nondecreasing if $r \to \log \beta_r^{1/(r-2)}$ is nondecreasing. Finally $r \to \log \beta_r^{1/(r-2)}$ is nondecreasing since

$$\log \beta_r^{1/(r-2)} = (r-2)^{-1} (\log \beta_r - \log \beta_2)$$

and the function $r \to \log \beta_r$ is convex on [2,m] (the convexity of the latter function is well known and follows immediately from Holder inequality).

From (35) - (36) it follows that (in the notations of (31),(34))

$$\begin{aligned} |\chi_{d_1} \cdots \chi_{d_v}| &\le c(d_1) \dots c(d_v)\beta_{j_1+2} \cdots \beta_{j_v+2} \\ &\le c(d_1) \dots c(d_v)\beta_2^{(r/2)+v} \frac{\beta_{j_1+2}}{\beta_2^{(j_1+2)/2}} \cdots \frac{\beta_{j_v+2}}{\beta_2^{(j_v+2)/2}} \end{aligned}$$

$$\leq c(d_1) \ldots c(d_v)\beta_2^{(r/2)+v}\,(\beta_m/\beta_2^{m/2})^{(j_1+\ldots+j_v)/(m-2)}$$

$$= c(d_1) \ldots c(d_v)\beta_2^{v-r/(m-2)}\beta_m^{r/(m-2)} ; \tag{37}$$

obviously

$$\beta_2^{v-r/(m-2)} \leq 1 + \beta_2^{r(m-3)/(m-2)} \tag{38}$$

On the other hand (since $d_1 + \ldots + d_v - a \geq 0$)

$$|t^{d_1+\ldots+d_v-a}| \leq |t|^{||d_1||+\ldots+||d_v||-||a||}$$

$$= |t|^{r+2v-||a||}$$

$$\leq 1 + |t|^{3r-||a||} \tag{39}$$

The lemma follows now from (34), (37) - (39). □

In what follows we shall need two auxiliary facts which we state as lemmas 6 and 7.

Lemma 6. Let f be a complex valued function defined on an open subset E of R^k. Suppose f has continuous derivatives D_b on E for all b with $||b|| \leq m$, where m is a positive integer.

1. If f has no zeros in E then $D_b \log f$, $||b|| \leq m$, exist and is given by

$$D_b \log f = \sum{}^* c(d_1,\ldots,d_j) \frac{D_{d_1} f \ldots D_{d_j} f}{f^j}$$

where the summation Σ^* is over all collections of nonzero nonnegative integral vectors $(d_1,\ldots,d_j)$ such that

$$\sum_1^j d_i = b,\ 1 \leq j \leq ||b||.$$

2. $D_b \exp\{f\}$ exists for all b with $||b|| \leq m$ and is given by

$$D_b \exp\{f\} = \exp\{f\} \sum c(\{j_d : 0 < d \leq b\})\ \Pi' (D_d f)^{j_d}$$

where the summation $\sum$ is over all collections of nonnegative integers $\{j_d : 0 < d \leq b\}$ such that $\sum_{0<d\leq b} d j_d = b$. The product Π' is taken over all $d : 0 < d \leq b$.

This lemma is easily proved by induction on b (if it is true for b then it is also true for $b+e_j$, $1 \leq j \leq k$, where e_j is the jth coordinate vector).

Denote by C^k the k-fold product of the complex plane C with itself and let $|\cdot|$ be the usual norm in C^k, i.e. $|z| = (\sum_1^k |z_j|^2)^{\frac{1}{2}}$ for $z = (z_1,\ldots,z_k) \in C^k$.

Lemma 7. Let f be an absolutely convergent power series on $B_r = \{z : z \in C^k, |z| < r\}$, such that

$$|f(z)| \leq h(|z|),\ z \in B_r\ ,$$

where h is a nondecreasing function on $[0,\infty)$. Then

$$|D_b f(z)| \leq b!(c|z|)^{-||b||}\ h((1+k^{\frac{1}{2}}\ c)|z|)$$

for any nonnegative integral vector b, $c > 0$ and $z \in B_{r_1}$, $r_1 = r(1+k^{\frac{1}{2}}\ c)^{-1}$.

<u>Proof</u>. Choose a number $c > 0$ and an element $z_o \in B_{r_1}$. Let $R = rk^{\frac{1}{2}}\ c(1+k^{\frac{1}{2}}\ c)^{-1}$. If $z \in B_R(z_o)$ (the ball of radius R with centre at z_o) then

$$\begin{aligned} |z| &\leq |z-z_o| + |z_o| \\ &< R + r_1 \\ &= r\ . \end{aligned}$$

Thus $B_R(z_o) \subset B_r$, $f(z)$ is defined on $B_R(z_o)$ and we have for $z \in B_R(z_o)$

$$f(z) = \sum c_d(z-z_o)^d, \tag{40}$$

where the summation is over all nonnegative integral vectors d. Consider a set

$$E_s = \{z : z\in C^k,\ |z_j-z_{oj}| = s, j=\overline{1,k}\},\ 0 < s < k^{-\frac{1}{2}}\ R.$$

If $z \in E_s$ then

$$|z-z_o| = k^{\frac{1}{2}}\ s < R,$$

hence $E_s \in B_R(z_o)$. The set E_s can be represented in the form

$$E_s = \{z : z \in C^k, z_j = z_{oj} + se^{i\theta_j},\ \theta_j \in [-\pi,\pi), j = \overline{1,k}\}.$$

Using this representation for E_s, multiplying (40) by $(2\pi)^{-k}$ $\exp\{-i(d',\theta)\}$ and integrating over E_s with respect to $d\theta_1 \ldots d\theta_k$ we obtain

$$(2\pi)^{-k} \int_{[-\pi,\pi)^k} e^{-i(d',\theta)}\ f(z_{01}+se^{i\theta_1},\ldots,z_{ok}+se^{i\theta_k})d\theta_1\ldots d\theta_k = c_{d'}s^{||d'||} \tag{41}$$

(here d' is a nonnegative integral vector). On the other hand taking b-th derivatives of the both sides of (40) at z_o we have

$$D_b\ f(z_o) = b!c_b. \tag{42}$$

Together (41) and (42) imply

$$|D_b f(z_o)| = |b!s^{-||b||}(2\pi)^{-k} \int_{[-\pi,\pi)^k} e^{-i(b,\theta)} f(z_{o1}+se^{i\theta_1},\dots,z_{ok}+se^{i\theta_k})d\theta_1..d\theta_k|$$

$$\leq b!\ s^{-||b||} \sup_{z\in E_s} \{|f(z)|\} . \qquad (43)$$

Take now $s = c|z_o|$ and observe that

$$E_{c|z_o|} \subset \{z:|z-z_o| = k^{\frac{1}{2}}c|z_o|\}$$

$$\subset \{z:|z| \leq |z-z_o| + |z_o| \leq (1+k^{\frac{1}{2}}c)|z_o|\}$$

$$\subset S_r .$$

Thus

$$\sup_{z\in E_{c|z_o|}} \{|f(z)|\} \leq \sup_{z\in E_{c|z_o|}} h(|z|)$$

$$\leq h((1+k^{\frac{1}{2}}c)\ |z_o|)$$

and the lemma follows. □

<u>Lemma 8</u>. Let X be a random variable with values in R^k such that $EX = 0, E|X|^m < \infty$ for some $m \geq 3$, and suppose that the covariance matrix V of X is nonsingular. Denote f the characteristic function of X, $\rho_s = E|V^{-\frac{1}{2}}X|^s, s > 0$, and let P_r be the polynomials defined by (30) with cumulants corresponding to X.

There exist constants $c(k,m)$, $\bar{c}(k,m)$ such that if

$$|t| < \bar{c}(k,m)n^{\frac{1}{2}}\ \rho_m^{-1/(m-2)} \qquad (44)$$

then

$$|D_a f^n(V^{-\frac{1}{2}}tn^{-\frac{1}{2}}) - \sum_{r=0}^{m-3} n^{-r/2}\ P_r(iV^{-\frac{1}{2}}t)e^{-|t|^2/2}|$$

$$\leq c(k,m)\rho_m\ n^{-(m-2)/2}(|t|^{m-||a||} + |t|^{3(m-2)+||a||})e^{-|t|^2/4}$$

for any nonnegative integral vector a with $||a|| \leq m$.

<u>Proof</u>. First assume that V=I. If $|t| < n^{\frac{1}{2}}\ \rho_m^{-1/(m-2)}$, then since $EX=0, |e^{i\alpha}-i\alpha-1| \leq \alpha^2/2$, and $\rho_m^{-1/m} \leq \rho_2^{-\frac{1}{2}} = k^{-1}$, we have

$$|f(tn^{-\frac{1}{2}}) - 1| = |E(\exp\{i(tn^{-\frac{1}{2}},X)\}-i(tn^{-\frac{1}{2}},X)-1|$$

$$\leq E(t,X)^2/2n$$

$$= |t|^2/2n$$

$$\leq 2^{-1}\ \rho_m^{-2/(m-2)}$$

$$\leq 2^{-1} k^{-2/(m-2)}$$

$$\leq 2^{-1}.$$

Thus assuming that $\bar{c} \leq 1$ in (44) we are in the domain where $\log f(tn^{-\frac{1}{2}})$ is well defined.

Denote now

$$g(t) = -|t|^2/2 + \sum_{r=1}^{m-3} n^{-r/2} \chi_{r+2}(it)/(r+2)!,$$

where $\chi_{r+2}(it)$ are defined by (30) with χ_a corresponding to X, and put

$$h(t) = n \log f(tn^{-\frac{1}{2}}) - g(t).$$

First we shall estimate

$$D_a(f^n(tn^{-\frac{1}{2}}) - \exp\{g(t)\})$$

$$= D_a(\exp\{h(t)\}-1).\exp\{g(t)\}$$

$$= \sum_{0 \leq b \leq a} c(a,b) D_{a-b}(\exp\{h(t)\}-1) D_b \exp\{g(t)\}. \tag{45}$$

Since V = I, g(t) may be written in the form

$$g(t) = \sum_{r=0}^{m-3} n^{-r/2} \chi_{r+2}(it)/(r+2)!$$

$$= -\sum_{r=0}^{m-3} n^{-r/2} i^r \sum_{||d||=r+2} (\chi_d/d!)t^d,$$

and we have for $b \geq 0$, $||b|| \leq m$ (since $|\chi_d| \leq c(d)\rho_{||d||}$ by (35))

$$|D_b g(t)| \leq | \sum_{r=\max(0,||b||-2)}^{m-3} n^{-r/2} \sum_{\substack{||d||=r+2 \\ d \geq b}} c'(d)\rho_{r+2}|t|^{r+2-||b||}$$

(by (36) and (44))

$$\leq \sum_{r=\max(0,||b||-2)}^{m-3} (\bar{c}(k,m))^r c(r)|t|^{2-||b||}$$

$$\leq c_1(k,m)|t|^{2-||b||}. \tag{46}$$

We shall use (46) for $b > 0$. When $b = 0$ we have in a similar way

$$|g(t) + |t|^2/2| \leq \sum_{r=1}^{m-3} (\bar{c}(k,m))^r c(r)|t|^2. \tag{47}$$

Choosing $\bar{c}(k,m)$ small enough the right side of (47) to be less than $|t|^2/8$ we shall have

$$|g(t)| \leq 5\,|t|^2/8\ . \tag{48}$$

Now if nonnegative integers $\{j_e : 0<e\leq b\}$ are such that

$$\textstyle\sum' e j_e = b,$$

$\sum'$ and Π' below denoting the summation and the product over all $e : 0 < e \leq b$ respectively, then

$$\begin{aligned}\Pi'(D_e g)^{j_e} &\leq c(k,m)|t|^{\Sigma' j_e(2-||e||)} \\ &\leq c(k,m)(|t|^{2-||b||} + |t|^{||b||}),\end{aligned}$$

since

$$1 \leq \textstyle\sum' j_e \leq \sum' j_e ||e|| = ||b||.$$

Noting now that by our choice of $\bar{c}(k,m)$ (cf. (47),(48))

$$\exp\{g(t)\} \leq \exp\{-(3/8)|t|^2\}, \tag{49}$$

we have by Lemma 6 for all $b > 0$, $||b|| \leq m$

$$\begin{aligned}|D_b \exp\{g(t)\}| &\leq c(k,m)\exp\{g(t)\}(|t|^{2-||b||} + |t|^{||b||}) \\ &\leq c(k,m)(|t|^{2-||b||} + |t|^{||b||})\exp\{-(3/8)|t|^2\}\ .\end{aligned} \tag{50}$$

If φ is a complex valued function with p continuous derivatives defined on R^k, then by Taylor's formula

$$\begin{aligned}&|\varphi(t) - \sum_{0\leq||d||\leq p-1} \frac{t^d}{d!} D_d\,\varphi(0)| \\ &\leq \sum_{||d||=p} \frac{|t^d|}{d!} \max\{|D_d\varphi(\theta t)|,\ 0 \leq \theta \leq 1\}\ .\end{aligned} \tag{51}$$

Noting that by the definition of cumulants

$$D_d h(0) = 0 \quad \text{for all } d : 0 \leq ||d|| \leq m-1$$

we have, applying (51) to $\varphi = D_\ell h$, $0 \leq ||\ell|| \leq m-1$ with $p = m-||\ell||$,

$$|D_\ell h(t)| \leq \sum_{||\ell'||=m-||\ell||} \frac{|t^{\ell'}|}{\ell'!} \sup\{|D_{\ell+\ell'} h(\theta t)|, 0 \leq \theta \leq 1\}$$

(obviously the last formula is also true when $||\ell|| = m$). Furthermore by Lemma 6 for ℓ with $||\ell|| = m$ we may write for any $t \in R^k$

$$|D_\ell h(t)| = n|D_\ell \log f(tn^{-\frac{1}{2}})|$$

$$\leq nc(k,m) \sum |D_{\ell_1} f(tn^{-\frac{1}{2}})| \ldots |D_{\ell_j} f(tn^{-\frac{1}{2}})|$$

where the summation is over all collections of nonzero nonnegative integral vectors $(\ell_1,\ldots,\ell_j)$ such that

$$\sum_1^j \ell_r = \ell,\ 1 \leq j \leq m$$

(as was noted above $|f(tn^{-\frac{1}{2}})| \geq \frac{1}{2}$ if (44) is satisfied with $\bar{c} \leq 1$). Since

$$|D_{\ell_r} f(tn^{-\frac{1}{2}})| \leq n^{-||\ell_r||/2} E|X^{\ell_r}|$$

$$\leq n^{-||\ell_r||/2} E|X|^{||\ell_r||}$$

$$\leq n^{-||\ell_r||/2} \rho_m^{||\ell_r||/m},$$

we obtain

$$|D_\ell h(t)| \leq c(k,m) \rho_m n^{-(m-2)/2}, \quad (||\ell|| = m),$$

and hence

$$|D_\ell h(t)| \leq c(k,m)\rho_m n^{-(m-2)/2} |t|^{m-||\ell||} (0 \leq ||\ell|| \leq m) \tag{52}$$

If $\ell = 0$ the condition (44) implies now that

$$|h(t)| \leq c(k,m)\rho_m n^{-(m-2)/2} |t|^m$$
$$\leq |t|^2/8 \tag{53}$$

when $\bar{c}(k,m)$ is small enough. Thus

$$|\exp\{h(t)\} - 1| \leq |h(t)| \exp\{|h(t)|\}$$
$$\leq c(k,m)\rho_m n^{-(m-2)/2} |t|^m \exp\{|t|^2/8\}. \tag{54}$$

For ℓ with $||\ell|| > 0$ we have by Lemma 6

$$D_\ell(\exp\{h(t)\}-1) = D_\ell \exp\{h(t)\}$$
$$= \exp\{h(t)\} \sum c(j_e{:}0 < e \leq \ell) \Pi'(D_e h)^{j_e}$$

where $\sum$ is the summation over all collections of nonnegative integers $\{j_e{:}0 < e \leq \ell\}$ such that

$$\sum_{0<e\leq\ell} ej_e = \ell$$

and Π' (and Σ' below) is the product (summation) over all $e{:}0 < e \leq \ell$. Applying

(52), using (44) and noting that

$$1 \le \sum{}' j_e \le ||\ell|| \le m$$

we have

$$\Pi'(D_e h)^{j_e} \le (c(k,m)\rho_m n^{-(m-2)/2})^{\Sigma' j_e} |t|^{\Sigma'(m-||e||)j_e}$$

$$\le c(k,m)\rho_m n^{-(m-2)/2}(|t|\rho_m^{1/(m-2)} n^{-\frac{1}{2}})^{(m-2)(\Sigma' j_e - 1)} \cdot$$

$$\cdot |t|^{m-2-||\ell||+2\Sigma' j_e}$$

$$\le c(k,m)\rho_m n^{-(m-2)/2}(|t|^{m-||\ell||} + |t|^{m-2+||\ell||}).$$

Together with (53) this imply

$$D_\ell(\exp\{h(t)\}-1) \le c(k,m)\rho_m n^{-(m-2)/2}(|t|^{m-||\ell||}+|t|^{m-2+||\ell||})\exp\{|t|^2/8\} \quad (55)$$

for all $\ell : 0 < ||\ell|| \le m$.

Relations (49), (54) when $a = 0$ and (50), (55) when $a \neq 0$ together with (45) give now

$$|D_a(f^n(tn^{-\frac{1}{2}}) - \exp\{g(t)\})|$$

$$\le c(k,m)\rho_m n^{-(m-2)/2}(|t|^{m-||a||}+|t|^{m+||a||})\exp\{-|t|^2/4\} \quad (56)$$

for all t satisfying (44).

The next step is to estimate

$$|D_a(\exp\{g(t)\} - \sum_{r=0}^{m-3} n^{-r/2} P_r(it)e^{-|t|^2/2})| = |D_a e(n^{-\frac{1}{2}}, it)e^{-|t|^2/2}| \quad (57)$$

$$= |\sum_{0 \le b \le a} c(a,b) D_{a-b}\, e^{-|t|^2/2}\, D_b e(n^{-\frac{1}{2}}, it)|,$$

where

$$e(n^{-\frac{1}{2}}, it) = \exp\{\sum_{r=1}^{m-3} n^{-r/2}\, \chi_{r+2}(it)/(r+2)!\} - \sum_{r=0}^{m-3} n^{-r/2} P_r(it).$$

By Lemma 6 we have for $0 \le b \le a, ||a|| \le m$

$$D_{a-b} e^{-|t|^2/2} \le c(k,m)(1+|t|^{||a-b||})\exp\{-|t|^2/2\}. \quad (58)$$

Consider now $e(y,z)$ obtained from $e(n^{-\frac{1}{2}}, it)$ by putting $y = n^{-\frac{1}{2}}$, $z = it$. We shall look at $e(y,z)$ assuming that

$$z \in C^k,\ u \in R^1,\ |y||z| \le \rho_m^{-1/(m-2)} \bar{c}(k,m) \quad (59)$$

where $\bar{c}(k,m)$ is the same as in (44) and is small enough (of course (59) coincide with

(44) when $y = n^{-\frac{1}{2}}$, $z = it$). We have

$$e(y,z) = \exp\{\sum_{r=1}^{m-3} y^r \frac{\chi_{r+2}(z)}{(r+2)!}\} - \sum_{r=0}^{m-3} y^r P_r(z)$$

and by the definition of polinomials $P_r(z)$

$$\frac{\partial^s}{\partial y^s} e(y,z)|_{y=0} = 0, \quad s = \overline{0,m-3}$$

$$\frac{\partial^{m-2}}{\partial y^{m-2}} e(y,z) = \frac{\partial^{m-2}}{\partial y^{m-2}} e_1(y,z),$$

where $e_1(y,z) = \exp e_2(y,z)$ and

$$e_2(y,z) = \sum_{r=1}^{m-3} y^r \frac{\chi_{r+2}(z)}{(r+2)!}$$

$$= \sum_{r=1}^{m-3} y^r \sum_{||\ell||=r+2} \frac{\chi_\ell}{\ell!} z^\ell \quad .$$

Thus

$$|e(y,z)| \leq \frac{|y|^{m-2}}{(m-2)!} \sup\{|\frac{\partial^{m-2}}{\partial y^{m-2}} e_1(\theta y,z)|, 0 \leq \theta \leq 1\} \quad .$$

By Lemma 6 the derivative $\frac{\partial^{m-2}}{\partial y^{m-2}} e_1(y,z)$ is a linear combination of terms of the form

$$e_1(y,z)(\frac{\partial}{\partial y} e_2(y,z))^{j_1} \ldots (\frac{\partial^{m-2}}{\partial y^{m-2}} e_2(y,z))^{j_{m-2}} \quad ,$$

where $j_1,\ldots,j_{m-2}$ are nonnegative integers such that

$$\sum_{s=1}^{m-2} s j_s = m-2 \quad .$$

Obviously

$$\frac{\partial^s}{\partial y^s} e_2(y,z) = 0 \text{ for } s = m-2$$

and for $0 \leq s \leq m-3$ by (35), (36) we have

$$|\frac{\partial^s}{\partial y^s} e_2(y,z)| = |\sum_{r=\max(1,s)}^{m-3} \frac{r!}{(r-s)!} (\sum_{||\ell||=r+2} \frac{z^\ell}{\ell!} \chi_\ell) y^{r-s}|$$

$$\leq c(k,m) \sum_{r=\max(1,s)}^{m-3} |z|^{r+2} \rho_{r+2} |y|^{r-s}$$

$$\leq c(k,m) \sum_{r=\max(1,s)}^{m-3} |z|^{r+2} \rho_m^{r/(m-2)} |y|^{r-s}$$

$$= c(k,m) \sum_{r=\max(1,s)}^{m-3} (|y|\,|z|\,\rho_m^{1/(m-2)})^{r-s} \rho_m^{s/m-2} |z|^{s+2} .$$

Now we shall use assumptions (59) to obtain

$$|e_2(y,z)| \le |z|^2/8$$

$$\left|\frac{\partial^s}{\partial y^s} e_2(y,z)\right| \le c(k,m)\ \rho_m^{s/(m-2)}|z|^{s+2} \quad (1 \le s \le m-2).$$

Thus

$$\left|\frac{\partial}{\partial y} e_2(y,z)\right|^{j_1} \cdots \left|\frac{\partial^{m-2}}{\partial y^{m-2}} e_2(y,z)\right|^{j_{m-2}}$$

$$\le\ c(k,m)\rho_m(|z|^m + |z|^{3(m-2)})$$

and so under the assumption (59) we have

$$|e(y,z)| \ \le\ c(k,m)\rho_m|y|^{m-2}(|z|^m + |z|^{3(m-2)})\exp\{|z|^2/8\}. \qquad (60)$$

To estimate derivatives of $e(y,z)$ with respect to z we shall apply Lemma 7 with $f(z) = e(y,z)$, $r = \frac{1}{|y|}\rho_m^{-1/(m-2)}\ \bar{c}(k,m)$, h equal to the right side of (60) and $c=3^{-1}k^{-\frac{1}{2}}$ to obtain

$$|D_b e(y,z)| \le c(k,m)\rho_m|y|^{m-2}(|z|^{m-||b||}+|z|^{3(m-2)-||b||})\exp\{(2/9)|z|^2\} \qquad (61)$$

for all $z{:}|z| \le (3/4)|y|^{-1}\ \rho_m^{-1/(m-2)}\ \ \bar{c}(k,m)$.

Combining now (57), (58) and (61) with $y = n^{-\frac{1}{2}}$, $z = it$ we have

$$\left|D_a(\exp\{g(t)\} - \sum_{r=0}^{m-3} n^{-r/2}\ P_r(it)e^{-|t|^2/2}\right|$$

$$\le\ c(k,m)\rho_m\ n^{-(m-2)/2}(|t|^{m-||a||} + |t|^{3(m-2)+||a||})\exp\{-|t|^2/4\} \quad (62)$$

for all t satisfying (44).

In the case $V=I$ the lemma follows now from (56) and (62).

If $V \ne I$ consider the random variable $X' = V^{-\frac{1}{2}}X$ which has mean zero and covariance matrix I. Denoting the characteristic function of X' by f' and the polynomials P_r corresponding to X' by P_r', we have $f'(t) = f(V^{-\frac{1}{2}}\, t)$, $P_r'(it) = P_r(iV^{-\frac{1}{2}}\, t)$. This implies the lemma in the general case. □

Proof of Theorem 1. Replacing if necessary X_j by $X_j-\mu$ we may suppose $\mu = 0$. We may also assume that

$$\beta_3\ n^{-\frac{1}{2}}\ \le\ (8k)^{-1} \qquad (63)$$

since otherwise

$$| \int f(x)(P_n-N)(dx)| \le 2\omega_f(R^k)$$

$$\le 16\, k\omega_f(R^k)\beta_3 n^{-\frac{1}{2}} .$$

Let Y_j, Z_j, $j = 1,\ldots,n$ be the random variables defined before Lemma 3, W be the covariance matrix of Z_1 and P'_n (resp. P''_n) be the distribution of $n^{-\frac{1}{2}} \sum_1^n Z_j$ (resp. $n^{-\frac{1}{2}} \sum_1^n Y_j$). We have

$$|\int f(x)(P_n-N)(dx)| \le | \int f(x)(P_n-P''_n)(dx)| + |\int f(x)(P''_n-N)(dx)| \tag{64}$$

and

$$|\int f(x)(P_n-P''_n)(dx)| = |E(f(n^{-\frac{1}{2}} \sum_1^n X_j)-f(n^{-\frac{1}{2}} \sum_1^n Y_j)|$$

$$\le \int_{\bigcup_1^n (X_j \neq Y_j)} |f(n^{-\frac{1}{2}} \sum_1^n X_j)-f(n^{-\frac{1}{2}} \sum_1^n Y_j)|dP$$

$$\le \omega_f (R^k)P(\bigcup_1^n (X_j \neq Y_j))$$

$$\le \omega_f(R^k)nP(|X_1| > n^{\frac{1}{2}})$$

$$\le \omega_f(R^k)\ \beta_3 n^{-\frac{1}{2}}. \tag{65}$$

Denote now $\alpha_n = n^{\frac{1}{2}} EY_1$ and let N'(resp. N") be normal $(0,W)$(resp. $(-\alpha_n,I)$) distribution. Then

$$|\int f(x)(P''_n-N)(dx)| = |\int f_{\alpha_n}(x)(P'_n-N'')(dx)|$$

$$\le |\int f_{\alpha_n}(x)(P'_n-N')(dx)|$$

$$+ |\int f_{\alpha_n}(x)(N'-N)(dx)|$$

$$+ |\int f_{\alpha_n}(x)|(N-N'')(dx)|$$

$$= I_1 + I_2 + I_3 . \tag{66}$$

By Lemma 3 we have

$$I_2 = | \int f_{\alpha_n}(x)(\tilde{\phi}(x)-\phi(x))dx|$$

$$\le \omega_f(R^k)\int|\tilde{\phi}(x)-\phi(x)|dx)$$

$$\le \omega_f(R^k)\ c(k)\beta_3\ n^{-\frac{1}{2}} \tag{67}$$

and

$$I_3 \leq \omega_f(R^k)\int|\phi(x)-\phi(x+\alpha_n)|dx$$

$$\leq \omega_f(R^k)\ c(k)\beta_3\ n^{-\frac{1}{2}} \tag{68}$$

It remains only to estimate I_1. By Lemma 1 there exist an absolutely continuous probability measure K on R^k such that

$$K(S_1(0)) \geq 3/4\ ,\ \int |x|^{k+1}\ K(dx) < \infty\ .$$

$$\hat{K}(t) = \int e^{i(t,x)}\ K(dx) = 0 \quad \text{if} \quad |t| > c_1(k).$$

Corollary 3 of § 2 implies now that for all $T > 0$

$$I_1 \leq 2[\omega_f(R^k)|(P_n'-N')*K_T| + \omega_f^*(2T^{-1};N')], \tag{69}$$

where $K_T(E) = K(TE)$ for any Borel set E and ω_f^* is defined by (21), § 2. In what follows we take

$$T = n^{\frac{1}{2}}/(16\ c_1(k)\beta_3),$$

and denote

$$J = |(P_n'-N')*K_T|\ .$$

Denoting p the density of $(P_n'-N')*K_T$ and applying Lemma 2 we may write

$$J = \int|p(x)|dx$$

$$\leq c(k) \max_{||a||=0,k+1} \int|D_a(\hat{P}_n'-\hat{N}')\hat{K}_T|dt$$

$$\leq c(k) \max_{||a+b||=0,k+1} \int|D_a(\hat{P}_n'-\hat{N}')D_b\hat{K}_T|dt\ .$$

Since (63) implies

$$T^{-1} = n^{-\frac{1}{2}}\ 16\ c_1(k)\beta_3$$

$$\leq 2c_1(k)k^{-1}\ ,$$

we have

$$|D_b\hat{K}_T(t)| = |\int D_b\ e^{i(t,x)}\ K_T(dx)|$$

$$= |\int D_b e^{i(t,x\ T^{-1})}\ K(dx)|$$

$$\leq T^{-||b||} \int|x^b|K(dx)$$

$$\leq (2c_1(k)k^{-1})^{||b||}\int|x|^{||b||}K(dx)$$

$$\leq c(k).$$

Thus

$$J \leq c(k) \max_{||a|| \leq k+1} \int_{|t| \leq A_n} |D_a(\hat{P}'_n - \hat{N}')| dt \tag{70}$$

where $A_n = n^{\frac{1}{2}}/16\ \beta_3$, since

$$\hat{K}_T(t) = \hat{K}(tT^{-1}) = 0 \text{ if } |t| > c_1(k)T = A_n .$$

Putting now

$$A'_n = (4/5)^{\frac{1}{2}}\ \bar{c}(k) n^{\frac{1}{2}}\ \rho_{k+2}^{-1/k} ,$$

where $\bar{c}(k) = \bar{c}(k,k+2)$ is from Lemma 8 and

$$\rho_{k+2} = E|W^{-\frac{1}{2}} Z_1|^{k+2}$$

we may write for any fixed a

$$\begin{aligned}\int_{|t| \leq A_n} |D_a(\hat{P}'_n - \hat{N}')| dt \leq & \int_{|t| \leq A'_n} |D_a(\hat{P}'_n - \sum_{r=0}^{k-1} n^{-r/2} P_r(it) e^{-(Wt,t)/2}| dt \\ & + \int |D_a \sum_{r=1}^{k-1} n^{-r/2} P_r(it) e^{-(Wt,t)/2}| dt \\ & + \int_{A'_n < |t| \leq A_n} |D_a \hat{P}'_n| dt \\ & + \int_{A'_n < |t| \leq A_n} |D_a \hat{N}'| dt \\ & = J_1 + J_2 + J_3 + J_4 ,\end{aligned} \tag{71}$$

where P_r are constructed by cumulants of Z_1. By (12) we have

$$\begin{aligned}|W^{-\frac{1}{2}}| & = |W^{-1}|^{\frac{1}{2}} \\ & \leq ||W^{-1}||^{k/2} \\ & \leq (4/3)^{k/2} ,\end{aligned} \tag{72}$$

and for any $u \in R^k$ by (11)

$$\begin{aligned}|W^{-\frac{1}{2}} u| & \geq ||W||^{-\frac{1}{2}} |u| \\ & \geq (4/5)^{\frac{1}{2}} |u| ,\end{aligned}$$

so that

$$\begin{aligned}\{u: |W^{-\frac{1}{2}} u| < A'_n\} & \subset \{u: |u| < (5/4)^{\frac{1}{2}} A'_n\} \\ & = \{u: |u| < \bar{c}(k) n^{\frac{1}{2}} \rho_{k+2}^{-1/k}\} .\end{aligned}$$

Thus putting $t = W^{-\frac{1}{2}} u$ in J_1 we have by Lemma 8

$$J_1 = \int_{|W^{-\frac{1}{2}} u| \leq A'_n} |D_a(\hat{P}'_n(W^{-\frac{1}{2}} u) - \sum_{r=0}^{k-1} n^{-r/2} P_r(iW^{-\frac{1}{2}} u) e^{-|u|^2/2})| |W^{-\frac{1}{2}}| du$$

$$\leq (4/3)^{k/2} \int\limits_{|u|\leq \bar{c}(k)n^{\frac{1}{2}}\rho_{k+2}^{-1/k}} |D_a(\hat{P}'_n(W^{-\frac{1}{2}} u) - \sum_{r=0}^{k-1} n^{-r/2} P_r(iW^{-\frac{1}{2}} u)e^{-|u|^2/2})|du$$

$$\leq c(k)\rho_{k+2}\, n^{-k/2} \quad .$$

Furthermore, by (72) and the relation (cf. (20))

$$\begin{aligned} E|Z_1|^q &\leq 2^q E|Y_1|^q \\ &\leq 2^q \beta_3 n^{(q-3)/2} \quad , \end{aligned} \tag{73}$$

which is valid for any $q \geq 3$, we have

$$\begin{aligned} \rho_{k+2} &= E|W^{-\frac{1}{2}}Z_1|^{k+2} \\ &\leq ||W^{-\frac{1}{2}}||^{k+2} E|Z_1|^{k+2} \\ &\leq c(k)\beta_3 n^{(k-1)/2}. \end{aligned} \tag{74}$$

Thus

$$J_1 \leq c(k)\beta_3 n^{-\frac{1}{2}} .$$

To estimate J_2 consider

$$\begin{aligned} L_r(t) &= D_a P_r(it)\exp\{-(Wt,t)/2\} \\ &= \sum_{0\leq b\leq a} c(a,b)D_b P_r(it)D_{a-b} \exp\{-(Wt,t)/2\} . \end{aligned} \tag{75}$$

By Lemma 5 with $m = r+2$

$$\begin{aligned} |D_b P_r(it)| &\leq c(k)(1+(E|Z_1|^2)^{r-1})(1+|t|^{3r-||b||})E|Z_1|^{r+2} , \\ &\quad \text{if} \quad ||b|| \leq 3r \\ &= 0, \text{ if } ||b|| > 3r. \end{aligned} \tag{76}$$

Note also that

$$\begin{aligned} E|Z_1|^2 &\leq 4E|Y_1|^2 \\ &\leq 4E|X_1|^2 \\ &= 4k . \end{aligned}$$

On the other hand, by (10)

$$\begin{aligned} \exp\{-(1/2)(Wt,t)\} &= \exp\{-|t|^2/2\} \exp \{(1/2)((I-W)t,t)\} \\ &\leq \exp \{-3|t|^2/8\}. \end{aligned} \tag{77}$$

Furthermore

$$D_\ell(-(1/2)(Wt,t)) = -\sum_{j=1}^{k} w_{ij}\, t_j \quad \text{for some } i \text{ if } ||\ell|| = 1$$

$$= -w_{ij} \quad \text{for some } i,j \text{ if } ||\ell|| = 2$$

$$= 0 \quad \text{if } ||\ell|| > 2$$

and by (9), (5)

$$|w_{ij} - \delta_{ij}| \leq 1/4k .$$

Thus applying Lemma 6 we have for any nonnegative integral vector e

$$D_e \exp\{-(Wt,t)/2\} \leq c(k)(1+|t|^{||e||})\exp\{-3|t|^2/8\} . \tag{78}$$

Relations (75)-(78) imply (in our case $||a|| \leq k+1$)

$$L_r(t) \leq c(k)(1+|t|^{3r+k+1}) \exp\{-3|t|^2/8\}E|Z_1|^{r+2}$$

which together with (73) for $q = r+2$ lead to the inequality

$$J_2 \leq c(k)\beta_3 n^{-\frac{1}{2}} .$$

Now using Lemma 4 and observing (77), (74) we have

$$J_3 \leq c(k) \int_{A_n' < |t| \leq A_n} (1+|t|^{||a||})\exp\{-5|t|^2/24\}|t|^k(A_n')^{-k}\, dt$$

$$\leq c(k)\rho_{k+2}\, n^{-k/2}$$

$$\leq c(k)\beta_3\, n^{-\frac{1}{2}}.$$

For J_4 we have by (78), (74)

$$J_4 \leq \int_{A_n' < |t| \leq A_n} |D_a \exp\{-(Wt,t)/2\}|t|^k(A_n')^{-k}\, dt$$

$$\leq c(k)\rho_{k+2}n^{-k/2}$$

$$\leq c(k)\beta_3 n^{-\frac{1}{2}} .$$

The obtained estimates for $J_1 - J_4$ give

$$J \leq c(k)\beta_3\, n^{-\frac{1}{2}} . \tag{79}$$

Finally by Lemma 3

$$\omega_f^*(2T^{-1};N') = \sup_y \int \omega_{f_y} (x,2T^{-1})\, \tilde{\phi}(x)dx$$

$$\leq \sup_y \int \omega_{f_y} (x,2T^{-1})\, \phi(x)dx$$

$$+ \sup_y \int \omega_{f_y} (x,2T^{-1})|\phi(x)-\tilde{\phi}(x)|dx$$

$$\leq \omega_f^*(2T^{-1},N) + c(k)\omega_f(R^k)\beta_3 n^{-\frac{1}{2}} \tag{80}$$

The theorem follows now from (64)-(69), (79) and (80). □

Theorem 1 and Theorem 1 of § 2 lead to the following result.

Theorem 2. Suppose the conditions of Theorem 1 are satisfied. Let ν and $\hat{\nu}$ be defined as in Theorem 1, § 2. Then there exist a constant $c(k)$ such that for any real bounded Borel measurable function f on R^k

$$|\int f(x)(P_n-N)(dx)| \leq c(k)[\omega_f(R^k)\hat{\nu}n^{-\frac{1}{2}} + \omega_f^*(\varepsilon_n',N)]$$

where $\varepsilon_n' = c(k)\nu n^{-\frac{1}{2}}$

Proof. If $\nu < c_1$ of Theorem 1, § 2 then

$$|\int f(x)(P_n - N)(dx)| \leq \omega_f(R^k)|P_n - N|(R^k)$$

$$\leq c_2\omega_f(R^k)\hat{\nu}n^{-\frac{1}{2}} .$$

On the other hand when $\nu \geq c_1$

$$\beta_3 \leq \nu + \int|x|^3 N(dx)$$

$$\leq (1 + cc_1^{-1}k^{3/2})\nu ,$$

and since the right side of (1) is nondecreasing in β_3 we may replace it by $(1 + cc_1^{-1}k^{3/2})\nu$.

Since $\nu \leq \hat{\nu}$, the theorem is proved. □

Remark 2. More general estimates of this type can be obtained by combining generalizations of Theorem 1, § 1 and of Theorem 1 (see § 2 Remark 3, [8],[40], [37], [38]).

Comments on Chapter I

The first estimate of the speed of convergence in the (one-dimensional) central limit theorem was obtained by A.M. Liapounov [21]. In his estimate in the i.i.d. case the bound was $c\beta_3 n^{-\frac{1}{2}} \log n$, i.e. differed from the Berry-Esseen bound only by the factor log n.

The Berry-Esseen theorem was proved at about the same time by A.C.Berry [6] and C.G. Esseen [14].

The method of characteristic functions in the theory of probability was introduced by A.M. Liapounov [20,21] and then developed by P.Levy [19] and others.

The method of compositions in the study of the accuracy of normal approximation was first used by H.Bergström [3-5] and elements of it may be traced back to Y.W. Lindeberg [22]. In [4] H. Bergström obtained the first multidimensional extension of the Berry-Esseen theorem (for distribution functions).

The indicated extensions and improvements 1-6 of the Berry-Esseen theorem belong respectively to M. Katz [17] (more precisely, an equivalent form of it belongs to M. Katz), S.V. Nagaev [23], V.V. Ulyanov [42] (a weaker version was obtained by V. Paulauskas [25]), R.N. Bhattacharya [7] (weaker forms were obtained by him earlier, see [8] for bibliography), C.G. Esseen [14], V.V. Sazonov [32].

Theorems 1 and 2 of § 2 were obtained by V.V. Sazonov and V.V. Ulyanov [37], [38]. The estimates of the constants in Theorem 1 § 1 presented here improve our previous estimates [38]. The present estimate for $\bar{c}(k)$ was first obtained by S. Nagaev (he used the method of characteristic functions) and then (together with other estimates) by V. Senatov [39]. After H. Bergstrom the method of compositions was developed by V.V. Sazonov [32, 34-36], V. Paulauskas [25-27], V.V. Ulyanov [41-42] and others. A method similar to the method of compositions was used in a very interesting paper by T.J. Sweeting [40]. However, due to the time limitations we could not include his work in our lectures. One of his main results states that in Theorem 1 of § 3 one can replace ω_f^* by $\bar{\omega}_f$. This result also permits to replace ω_f^* by $\bar{\omega}_f$ in Theorem 2 of § 3.

Theorem 1 of § 3 belongs to R.N. Bhattacharya [7]. Our presentation follows R.N. Bhattacharya and R. Ranga Rao book [8] but is simpler due to the fact that we consider a less general case. We also hope that we made some points of the proof more clear. In the general case considered in [8] X_j may not be identically distributed and the function f may be unbounded (with some restrictions on its growth at infinity; see also [40]). The progress in the estimation of the speed of convergence in the multidimensional central limit theorem based on the method of characteristic functions was made by C.G. Esseen [15], R. Ranga Rao [29], B. Von Bahr [1,2], R.N. Bhattacharya [7-9], V.I. Rotar [30,31], A. Bikelis [10] and others.

Theorem 2 of § 3 belongs to V.V. Sazonov and V.V. Ulyanov [37].

CHAPTER II

THE INFINITE DIMENSIONAL CASE

In this chapter we shall prove two theorems on the speed of convergence in the central limit theorem in Hilbert space. The situation here differs substantially from the finite dimensional case and our knowledge in this area is less advanced.

Let $X_1, X_2, \ldots$ be a sequence of independent random variables with values in a real separable Hilbert space H and with the same distribution P. Suppose that $E|X_1|^2 < \infty$, $EX_1 = \mu$. EX_1 is understood in the sense of Bochner or Pettis integral, under the assumption $E|X_1|^2 < \infty$ they both exist and coincide and denote V the covariance operator corresponding to P, i.e. the operator defined by

$$(Vx,x) = \int (x,z-\mu)(y,z-\mu)P(dz) .$$

The central limit theorem in H states that the distributions P_n of the normalized sums $n^{-\frac{1}{2}} \sum_1^n (X_r-\mu)$ converge weakly to the normal distribution G with mean zero and covariance operator V, i.e. the distribution defined by

$$\int \exp\{i(x,y)\}G(dy) = \exp\{-(1/2)(Vx,x)\} .$$

Note that contrary to the finite dimensional case even if E is a rather small class of "good" subsets of H (e.g. the class of all halfspaces, i.e. sets of the form $\{x:(x,y) < u\}$, $y \in H$, $u \in R$, or the class of all balls) the convergence

$$P_n(E) \to G(E), \; E \in E \tag{1}$$

implied by the central limit theorem may not be uniform in $E \in E$. Indeed let $\{e_s\}$, $s = 1,2,\ldots$ be an orthonormal basis in H and X_{rs}, $r,s = 1,2,\ldots$ be independent real random variables such that

$$X_{rs} = \begin{cases} 0 & \text{with probability } p_s \\ \pm\ell_s & \text{with probability } q_s \end{cases} \qquad (r,s = 1,2,\ldots)$$

where

$$0 < q_s < 1, \; p_s + 2q_s = 1, \; q_s \to 0 \text{ as } s \to \infty, \; \sum_{s=1}^{\infty} \ell_s^2 = L^2 < \infty. \text{ Put}$$

$X_r = \sum_{s=1} X_{rs} e_s$. Obviously X_r are i.i.d, $EX_1 = 0$ and $|X_1| \le L$ so that $E|X_1|^2 < \infty$.

Take E to be the class E_1 of all halfspaces. We have for any fixed n, $\varepsilon > 0$

$$P_n(x:(e_s,x) = 0) = P((e_s, n^{-\frac{1}{2}} \sum_{r=1}^{n} X_r) = 0)$$

$$= P(n^{-\frac{1}{2}} \sum_{r=1}^{n} X_{rs} = 0)$$

$$\geq \prod_{r=1}^{n} P(X_{rs} = 0)$$

$$= p_s^n$$

$$> 1 - \varepsilon$$

for all large enough s (say s S). At the same time

$$G(x:(e_s,x) = 0) = 0$$

and we obtain for any $s \geq S$

$$\sup_{E \in E_1} |P_n(E) - G(E)| \geq \sup_{u \in R} |P_n(x:(e_s,x) < u) - G(x:(e_s,x) < u)|$$

$$\geq (1/2) P_n(x:(e_s,x) = 0)$$

$$\geq (1/2)(1 - \varepsilon) .$$

Since $\varepsilon > 0$ is arbitrary this implies

$$\sup_{E \in E_1} |P_n(E) - G(E)| \geq 1/2 \tag{2}$$

Inequality (2) still holds if we replace E_1 by the class E_2 of all balls in H. This follows immediately from (2) if we observe that for any s and u the balls $\{x: |x + (R - u)e_s| < R\}$ increase with R and their union is the halfspace $\{x:(e_s,x) < u\}$. However, if we take E to be the class of balls with a fixed center then the convergence (1) is uniform in $E \in E$ and we shall estimate the speed of this convergence.

§ 1. An estimate for balls with a fixed center

Theorem 1. Let $X_1, X_2, \ldots$ be a sequence of independent random variables with values in a separable Hilbert space H with the same distribution P. Suppose that $E|X|^2 < \infty$ and that the covariance operator V of P is injective. Denote P_n the distribution of the normalized sum $n^{-\frac{1}{2}} \sum_1^n (X_i - \mu)$ where $\mu = EX_1$ and let G be normal $(0,V)$ distribution. Then for any $h \in H$, $n \geq 2$

$$\Delta_n(h) = \sup_{r\geq 0} |P_n(S_r(h)) - G(S_r(h))| \leq \bar{c}\bar{c}(V)((trV)^{\frac{1}{2}} + |h|)\nu^{1/3} n^{-1/6},$$

where

$$\nu = \int |x|^3 |P_{X_1-\mu} - G|(dx), \ \bar{c}(V) = \sigma_1^{-1}\sigma_2^{-1} \prod_{j=3}^{\infty} (1 - \sigma_j^2/2\sigma_3^2)^{-\frac{1}{2}}.$$

The proof of this theorem is based on Lemmas 1-5.

Lemma 1. Let G be normal $(0,V)$ distribution on H with V injective. There exists a constant c such that for any $r > 0$, $\varepsilon \geq 0$, $h \in H$

$$G(\bar{S}_{r,\varepsilon}(h)) \leq c\bar{c}(V)(\sigma_1 + |h|)\varepsilon \tag{3}$$

Proof. Let $\sigma_1^2 \geq \sigma_2^2 \geq \ldots$ be eigenvalues and $e_1, e_2, \ldots$ be corresponding eigenvectors of $V(e_1, e_2, \ldots$ constitute a basis in $H)$. Then G is the distribution of the random variable $Y = \sum_{j=1}^{\infty} \sigma_j Y_j e_j$ where $\{Y_j\}$ are i.i.d. normal $(0,1)$ real random variables. In particular

$$G(\bar{S}_{r,\varepsilon}(h)) = P(r \leq |Y - h| \leq r + \varepsilon) \ .$$

Denote f_j, g_1, g_2, g, f the densities corresponding to the random variables $\bar{Y}_j = (\sigma_j Y_j - h_j)^2$, $\bar{Y}_1 + \bar{Y}_2$, $\sum_{j=3}^{\infty} \bar{Y}_j$, $\sum_{j=1}^{\infty} \bar{Y}_j = |Y-h|^2, |Y-h|$ respectively, where $h_j = (h, e_j)$. It is easy to calculate (Y_j being standard normal) that

$$f_j(u) = \frac{u^{-\frac{1}{2}}}{(2\pi)^{\frac{1}{2}}\sigma_j} \left[\frac{1}{2}\left(\exp\left\{-\frac{(u^{\frac{1}{2}}+h_j)^2}{s\sigma_j^2}\right\} + \exp\left\{-\frac{(-u^{\frac{1}{2}}+h_j)^2}{2\sigma_j^2}\right\}\right)\right] . \tag{4}$$

Denoting the expression in the square brackets in (4) by $d_j(u)$ we may write

$$\begin{aligned} g_1(u) &= \int_0^u f_1(u-v) f_2(v) dv \\ &= \frac{1}{2\pi\sigma_1\sigma_2} \int_0^u \frac{d_1(u-v) d_2(v)}{(u-v)^{\frac{1}{2}} v^{\frac{1}{2}}} dv \\ &\leq \frac{1}{2\pi\sigma_1\sigma_2} \int_0^1 \frac{dw}{(1-w)^{\frac{1}{2}} w^{\frac{1}{2}}} \\ &= c\sigma_1^{-1}\sigma_2^{-1} . \end{aligned} \tag{5}$$

On the other hand if $u \geq 4h'$, $h' = h_1^2 + h_2^2$, $w \geq \frac{1}{2}$, then since $(\alpha+\beta)^{\frac{1}{2}} \geq 2^{-\frac{1}{2}}(\alpha^{\frac{1}{2}} + \beta^{\frac{1}{2}})$, $\alpha, \beta \geq 0$

$$\begin{aligned} (uw)^{\frac{1}{2}} \pm h_j &\geq 2^{-\frac{1}{2}}(u-4h')^{\frac{1}{2}} w^{\frac{1}{2}} + 2^{\frac{1}{2}}(h')^{\frac{1}{2}} w^{\frac{1}{2}} \pm h_j \\ &\geq 2^{-1}(u-4h')^{\frac{1}{2}}, \ j = 1, 2, \end{aligned}$$

and we have

$$d_j(uw) \le \exp\left\{-\frac{u-4h'}{8\sigma_j^2}\right\}, \quad j = 1,2 .$$

Thus

$$g_1(u) = \frac{1}{2\pi\sigma_1\sigma_2}\int_0^1 \frac{d_1(u(1-w))d_2(uw)}{(1-w)^{\frac{1}{2}}w^{\frac{1}{2}}}\,dw$$

$$\le \frac{c}{\sigma_1\sigma_2}\exp\left\{-\frac{u-4h'}{8\sigma_1^2}\right\} \tag{6}$$

for all $u \ge 4h'$.

Furthermore since

$$g(u) = \int_0^u g_1(u-v)g_2(v)\,dv$$

we have by (5)

$$g(u) \le c\sigma_1^{-1}\sigma_2^{-2} . \tag{7}$$

On the other hand

$$g(u) = \left(\int_0^{u/2} + \int_{u/2}^{u}\right) g_1(u-v)g_2(v)\,dv$$

$$= I_1 + I_2 ,$$

and by (6)

$$I_1 \le \frac{c}{\sigma_1\sigma_2}\exp\left\{-\frac{1}{8\sigma_1^2}\left(\frac{u}{2} - 4h'\right)\right\}$$

$$\le \frac{c}{\sigma_1\sigma_2}\exp\left\{-\frac{1}{8\sigma_1^2}\left(\left(\frac{u}{2}\right)^{\frac{1}{2}} - 2(h')^{\frac{1}{2}}\right)^2\right\}$$

for all $u \ge 8h'$ since $\alpha^2 - \beta^2 \ge (\alpha-\beta)^2$ for $\alpha \ge \beta \ge 0$.

Note now that if $\tilde{Y} = 2\sigma_3^2(Y_3^2 + Y_4^2)$, $\hat{Y} = \sum_{j=5}^{\infty}\sigma_j^2 Y_j^2$ then for any $w \ge 0$

$$P\left(\sum_{j=3}^{\infty}\sigma_j^2 Y_j^2 \ge w\right) \le P(\tilde{Y} + \hat{Y} \ge w)$$

$$= \int_0^{\infty} P(\tilde{Y} \ge w - s)P_{\hat{Y}}(ds)$$

$$= e^{-w/4\sigma_3^2}\int_0^{\infty} e^{s/4\sigma_3^2} P_{\hat{Y}}(ds)$$

$$\le c_1(V)e^{-w/4\sigma_3^2}, \quad c_1(V) = \prod_{j=5}^{\infty}(1 - \sigma_j^2/2\sigma_3^2)^{-\frac{1}{2}},$$

since $2^{-1}\sigma_3^{-2}\tilde{Y}$ is χ^2 random variable with two degrees of freedom. Consequently

$$P\left(\sum_{j=3}^{\infty}\bar{Y}_j \ge u/2\right) \le P\left(\left(\sum_{j=3}^{\infty}\sigma_j^2 Y_j^2\right)^{\frac{1}{2}} \ge (u/2)^{\frac{1}{2}} - (h'')^{\frac{1}{2}}\right)$$

$$\le c_1(V)\exp\{-(1/4\sigma_3^2)((u/2)^{\frac{1}{2}} - (h'')^{\frac{1}{2}})^2\}$$

for all $u \geq 2h''$, where $h'' = \sum_{j=3}^{\infty} h_j^2$ and we obtain for such u

$$I_2 \leq c\bar{c}(V) \exp\{-(1/4\sigma_3^2)((u/2)^{\frac{1}{2}} - (h'')^{\frac{1}{2}})^2\} .$$

Thus for all $u \geq 8|h|^2$

$$g(u) \leq c\bar{c}(V) \exp\{-(1/8\sigma_1^2)((u/2)^{\frac{1}{2}} - 2|h|)^2)\} . \tag{8}$$

Finally by (7) and (8)

$$\begin{aligned} f(u) &= 2\, ug(u^2) \\ &\leq \sup_{0 \leq u \leq 8|h|^2} 2u^{\frac{1}{2}}h(u) + \sup_{8|h|^2 \leq u} 2u^{\frac{1}{2}}g(u) \\ &\leq c\bar{c}(V)(\sigma_1 + |h|) . \end{aligned}$$

which implies (3). □

Lemma 2. Let f be a real nonincreasing function on R such that

$$\begin{aligned} f(u) &= 1 \quad \text{if} \quad u \leq 0 \\ &= 0 \quad \text{if} \quad u \geq \varepsilon \end{aligned}$$

for some $\varepsilon > 0$. Then for any two Borel probability measures Q_1, Q_2 on H

$$\begin{aligned} \sup_{r \geq 0} |Q_1(S_r(h)) - Q_2(S_r(h))| &\leq \sup_{s>0} |\int f(|x-h|-s)(Q_1 - Q_2)(dx) \\ &\quad + 3 \sup_{s \geq 0} Q_2(\bar{S}_{s,\varepsilon}(h)) . \end{aligned}$$

Proof. Denote for brevity $R = Q_1 - Q_2$. If $R(S_r(h)) \geq 0$ then since

$$\chi_{S_r(h)} \leq f(|x-h| - r)$$

$$f(|x-h| - r) - \chi_{S_r(h)} \leq \chi_{S_{r,\varepsilon}(h)}$$

we have

$$\begin{aligned} |R(S_r(h))| &= \int \chi_{S_r(h)}(x) R(dx) \\ &\leq \int f(|x-h| - r) R(dx) + Q_2(S_{r,\varepsilon}(h)) . \end{aligned} \tag{9}$$

If $R(S_r(h)) < 0$ then since

$$\chi_{S_r^c(h)} \leq 1 - f(|x-h| - r + \varepsilon)$$

$$1 - f(|x-h| - r + \varepsilon) - \chi_{S_r^c(h)} \leq \chi_{S_{r-\varepsilon,\varepsilon}}(h)$$

we have

$$|R(S_r(h))| = R(S_r^c(h))$$
$$\leq \int(1 - f(|x-h| - r + \varepsilon))R(dx) + Q_2(S_{r-\varepsilon,\varepsilon}(h))$$
$$\leq |\int f(|x-h| - r + \varepsilon)R(dx)| + Q_2(S_{r-\varepsilon,\varepsilon}(h)) \ . \tag{10}$$

Note also that if $t \leq 0$, $\delta > 0$, then since

$$f(|x-h| - t) \leq f(|x-h| - \delta)$$
$$\leq \chi_{S_{\varepsilon+\delta}}(h)$$
$$f(|x-h| - t) \leq \chi_{S_\varepsilon}(h)$$

we have

$$|\int f(|x-h| - t)R(dx)| \leq \int f(|x-h| - t)(Q_1 + Q_2)(dx)$$
$$\leq \int f(|x-h| - \delta)R(dx) + Q_2(S_\varepsilon(h)) + Q_2(S_{\varepsilon+\delta}(h))$$

and since $\delta > 0$ is arbitrary it follows that

$$|\int f(|x-h| - t)R(dx)| \leq \sup_{s<0} |\int f(|x-h| - s)R(dx)| + 2Q_2(\bar{S}_\varepsilon(h)). \tag{11}$$

Now (9) - (11) imply for any $r > 0$

$$|R(S_r(h))| \leq \sup_{s>0} |\int f(|x-h| - s)R(dx)| + 3 \sup_{s\geq 0} Q_2(\bar{S}_{s,\varepsilon}(h)).$$

Letting $r \to 0$ we see that this inequality is also true for $r = 0$ and the lemma follows. □

For each $\varepsilon > 0$ let f_ε be a real non-increasing function on R with three continuous derivatives such that

$$f_\varepsilon(u) = 1 \quad \text{if} \quad u \leq 0$$
$$= 0 \quad \text{if} \quad u \geq \varepsilon \tag{12}$$
$$|f_\varepsilon^{(j)}(u)| \leq c\varepsilon^j, \ j = 1,2,3.$$

For example we can take $f_\varepsilon(u) = p(u\varepsilon^{-1})$ where

$$p(u) = 1 \quad \text{if} \quad u \leq 0$$
$$= (1 - u^4)^4 \quad \text{if} \quad 0 \leq u \leq 1$$
$$= 0 \quad \text{if} \quad u \geq 1.$$

For fixed z_1, $z \in H$, $s > 0$ define

$$g_\varepsilon(t) = f_\varepsilon(|z_1 + tz| - s), t \in R.$$

Lemma 3. $g_\varepsilon(t)$ is three times continuous differentiable and

$$|g_\varepsilon^{(j)}(t)| \le c\varepsilon^{-j}|z|^j, \; j = 1,2,3 \tag{13}$$

Proof. Denote $h(t) = |z_1 + tz|$. For t with $h(t) = 0$ (13) being obvious we may suppose $h(t) \neq 0$. We have

$$\begin{aligned} g_\varepsilon'(t) &= f_\varepsilon'(h(t) - s)h'(t) \\ g_\varepsilon''(t) &= f_\varepsilon''(h(t) - s)(h'(t))^2 + f_\varepsilon'(h(t) - s)h''(t) \\ g_\varepsilon''(t) &= f_\varepsilon'''(h(t) - s)(h'(t))^3 + 3f_\varepsilon''(h(t) - s)h'(t)h''(t) + f_\varepsilon'(h(t)-s)h'''(t) \end{aligned} \tag{14}$$

and a straightforward calculation gives

$$\begin{aligned} h'(t) &= \frac{(z_1 + tz, z)}{h(t)} \\ h''(t) &= \frac{|z|^2}{h(t)} - \frac{(z_1 + tz, z)^2}{h(t)^3} \\ h'''(t) &= -3\frac{(z_1 + tz, z)|z|^2}{h(t)^3} + 3\frac{(z_1 + tz, z)^3}{h(t)^5}. \end{aligned} \tag{15}$$

Note now that $f_\varepsilon(v - s)$ as a function of $v \in R$ is three times continuously differentiable and $f_\varepsilon^{(j)}(-s) = 0$, $j = 1,2,3$. Thus

$$\begin{aligned} f_\varepsilon'(v - s) &= \int_o^v f_\varepsilon''(s_1 - s)ds_1 \\ &= \int_o^v \int_o^{s_1} f_\varepsilon'''(s_2 - s)ds_2 ds_1 \end{aligned}$$

and we obtain for $v \ge 0$

$$|f_\varepsilon^{(i)}(v - s)| \le v^{j-i} \sup_u |f_\varepsilon^{(j)}(u)|, \; 1 \le i \le j \le 3. \tag{16}$$

The Lemma follows now immediately from (12) - (16). □

Lemma 4. Let G be normal $(0,V)$ distribution on H and P be a probability distribution on H with the same mean and covariance operator as G. For $\alpha > 0$ define distributions $P_{(\alpha)}, G_\alpha$ by

$$P_{(\alpha)}(B) = P(\alpha B), G_\alpha(B) = G(\alpha B) \tag{17}$$

where B are Borel sets. Then for any $h \in H$, $s,\alpha,\beta,\gamma > 0$ and $i = 0,1,2$

$$|\int f_\varepsilon(|x-h|-s)P^i_{(\alpha)}*G_\beta*Q_\gamma(dx)| \le c(1+\varepsilon^{-1}+\varepsilon^{-2})\gamma^{-3}\nu_1$$
$$+ c\bar{c}(V)[\sigma_1+\beta(1+\alpha^{-1}(i\ \mathrm{tr}\ V)^{1/2}+|h|)]\beta\varepsilon^{-2}\gamma^{-3}\nu_1 ,$$

where f_ε is defined in (12), $Q_\gamma = P_{(\gamma)} - G_\gamma$,

$$\nu_1 = \int |x|^3 \, |Q_1|(dx) .$$

<u>Proof</u>. For any $x,y,z \in H$ denote

$$\bar{g}_\varepsilon(t) = \bar{g}_{\varepsilon,x,y,z}(t) = f_\varepsilon(|x+y+tz-h|-s) . \tag{18}$$

The properties of f_ε imply obviously

$$\bar{g}^{(j)}_\varepsilon(t) = \bar{g}^{(j)}_\varepsilon(t)\chi_{S_{s,\varepsilon}(h-x-tz)}(y),\ j=1,2,\ldots, \tag{19}$$

and

$$\bar{g}^{(j)}_\varepsilon(0) = 0 \quad \text{if} \quad |x+y-h| = 0,\ j=1,2,\ldots .$$

For $|x+y-h| \ne 0$ denoting $z_1 = x+y-h$ we have by (14), (15)

$$\bar{g}'_\varepsilon(0) = f'_\varepsilon(|z_1|-s)\frac{(z_1,z)}{|z_1|}$$
$$\bar{g}''_\varepsilon(0) = f''_\varepsilon(|z_1|-s)\frac{(z_1,z)^2}{|z_1|^2} + f'_\varepsilon(|z_1|-s)\left(\frac{|z|^2}{|z_1|} - \frac{(z_1,z)^2}{|z_1|^3}\right) . \tag{20}$$

Observe that both P_γ, G_γ have mean zero and covariance operator $\gamma^{-2}V$. Thus for any $z' \in H$

$$\int (z',z)Q_\gamma(dz) = 0$$
$$\int (z',z)^2 Q_\gamma(dz) = 0$$
$$\int |z|^2 \, Q_\gamma(dz) = 0$$

and we see that for any $x,\ y \in H$

$$\int \bar{g}^{(j)}_{\varepsilon,x,y,z}(0)Q_\gamma(dz) = 0,\ j=0,1,2 .$$

Consequently we may write

$$|\int f_\varepsilon(|x-h|-s)P^i_{(\alpha)}*G_\beta*Q_\gamma(dz)| = |\iiint \bar{g}_\varepsilon(1)P^i_{(\alpha)}(dx)G_\beta(dy)Q_\gamma(dz)|$$
$$= |\iiint(\bar{g}_\varepsilon(1) - \sum_{j=0}^{2}\bar{g}^{(j)}_\varepsilon(0)/j!)P^i_{(\alpha)}(dx)G_\beta(dy)\times Q_\gamma(dz)$$
$$\le \sum_1^5 I_r . \tag{21}$$

where

$$I_1 = \frac{1}{6} \iiint_{z\in S_1} |\bar{g}_\varepsilon'''(\theta)| P^i_{(\alpha)}(dx)\, G_\beta(dy)\, |Q_\gamma|(dz)$$

$$I_2 = \iiint_{z\in S_1^c} \bar{g}_\varepsilon(1) P^i_{(\alpha)}(dx)\, G_\beta(dy)\, |Q_\gamma|(dz)$$

$$I_{3+j} = \iiint_{z\in S_1^c} |\bar{g}_\varepsilon^{(j)}(0)| P^i_{(\alpha)}(dx)\, G_\beta(dy)\, |Q_\gamma|(dz), \quad j=0,1,2 .$$

To estimate $I_2 - I_5$ we use Lemma 3 and the inequality

$$\int_{S_1^c} |z|^j |Q_\gamma|(dz) \le \gamma^{-3} \nu_1, \quad j=0,1,2$$

to obtain

$$\sum_2^5 I_r \le c\gamma^{-3}(1+\varepsilon^{-1}+\varepsilon^{-2})\nu_1 . \tag{22}$$

Furthermore by (19) and Lemma 3

$$I_1 \le c\varepsilon^{-3} \int_{S_1} |z|^3 |Q_\gamma|(dz) \iint \chi_{S_{s,\varepsilon}(h-x-\theta z)}(y) P^i_{(\alpha)}(dx)\, G_\beta(dy) . \tag{23}$$

Now by Lemma 1

$$G_\beta(S_{s,\varepsilon}(h-x-\theta z)) = G(S_{\beta s,\beta\varepsilon}(\beta(h-x-\theta z)))$$
$$\le c\bar{c}(V)(\sigma_1 + \beta(1+|x|+|h|)\beta\varepsilon ,$$

and since

$$\int |x| P^i_{(\alpha)}(dx) \le \alpha^{-1}\left(\int |x|^2 P^i_{(\alpha)}(dx)\right)^{\frac{1}{2}}$$
$$= \alpha^{-1}(i \ \mathrm{tr}\ V)^{\frac{1}{2}} ,$$

we have

$$\iint \chi_{S_{s,\varepsilon}(h-x-\theta z)}(y) P^i_{(\alpha)}(dx)\, G_\beta(dy) \le c\bar{c}(V)(\sigma_1 + \beta(1+\alpha^{-1}(i \ \mathrm{tr}\ V)^{\frac{1}{2}} + |h|))\beta\varepsilon .$$

Thus

$$I_1 \le c\bar{c}(V)[\sigma_1 + \beta(1+\alpha^{-1}(i \ \mathrm{tr}\ V)^{\frac{1}{2}} + |h|]\beta\gamma^{-3}\varepsilon^{-2}\nu_1 . \tag{24}$$

The lemma follows now from (21) - (24). □

<u>Lemma 5</u>. If P is any probability distribution on H, G is normal $(0,V)$ distribution on H and $Q = P - G$ then

$$|Q|(H) \le c(\sigma_1 \bar{c}(V))^{3/4} \nu_1^{1/4} \tag{25}$$

where $\nu_1 = \int |x|^3 \, |Q|(dx)$. If moreover P has mean zero and covariance operator V then for all $n \ge 2$,

$$|P_n(S_r(h)) - G(S_r(h))| \le c(n)(1+\bar{c}(V))[1+(\mathrm{tr}\ V)^{\frac{1}{2}} + |h|]\nu_1^{1/3} \tag{26}$$

where P_n is defined by $P_n(B) = P^n(n^{\frac{1}{2}}B)$.

Proof. To prove (25) we note that for any $\varepsilon > 0$

$$\begin{aligned}|Q|(S_\varepsilon) &\le (P+G)(S_\varepsilon)\\ &= -Q(S_\varepsilon^c) + 2G(S_\varepsilon)\\ &\le |Q|(S_\varepsilon^c) + 2G(S_\varepsilon)\\ |Q|(S_\varepsilon^c) &\le \varepsilon^{-3}\int_{S_\varepsilon^c}|x|^3|Q|(dx)\\ &\le \varepsilon^{-3}\nu_1 .\end{aligned}$$

Moreover by Lemma 1

$$G(S_\varepsilon) \le c\sigma_1\bar{c}(V)\varepsilon .$$

Thus

$$|Q|(H) \le 2(\varepsilon^{-3}\nu_1 + c\sigma_1\bar{c}(V)\varepsilon) .$$

Putting here $\varepsilon = (\sigma_1\bar{c}(V))^{-\frac{1}{4}}\nu_1^{\frac{1}{4}}$ we obtain (25).

To prove (26) observe that by Lemmas 1,2

$$|P_n(S_r(h)) - G(S_r(h))| \le \sup_{s>0} |\int f_\varepsilon(|x-h|-s)(P_n-G)(dx)| + c\bar{c}(V)(\sigma_1+|h|)\varepsilon . \tag{27}$$

Furthermore defining $P_{(\alpha)}$, G_α, $\alpha > 0$ as in (17) and putting $Q_\alpha = P_{(\alpha)} - G_\alpha$ we may write

$$\begin{aligned}P_n - G &= P^n_{(n^{\frac{1}{2}})} - G^n_{n^{\frac{1}{2}}}\\ &= \sum_{i=1}^{n}\binom{n}{i} Q^i_{n^{\frac{1}{2}}} * G^{n-i}_{n^{\frac{1}{2}}}\\ &= Q^n_{n^{\frac{1}{2}}} + \sum_{i=0}^{n-2} c(i,n)Q_{n^{\frac{1}{2}}} * P^i_{(n^{\frac{1}{2}})} * G^{n-i-1}_{n^{\frac{1}{2}}} .\end{aligned} \tag{28}$$

If f is any real measurable function on H with $0 \le f \le 1$ then

$$|\int f(x)Q^n_{n^{\frac{1}{2}}}(dx)| \le (1/2)[|Q|(H)]^n . \tag{29}$$

Indeed (29) is obvious for $n = 1$ and for $n > 1$ using induction on n we have

$$\begin{aligned}|\int f(x)Q^n_{n^{\frac{1}{2}}}(dx)| &= |\int\int f(x+y)Q^{n-1}_{n^{\frac{1}{2}}}(dx)Q_{n^{\frac{1}{2}}}(dy)|\\ &\le (1/2)[|Q|(H)]^{n-1}|Q_{n^{\frac{1}{2}}}|(H)\\ &= (1/2)[|Q|(H)]^n .\end{aligned}$$

On the other hand, since $G^{n-i-1}_{n^{1/2}} = G_{(n/(n-i-1))^{1/2}}$, by Lemma 4 with $\alpha = n^{1/2}$, $\beta = (n/(n-i-1))^{1/2}$, $\gamma = n^{1/2}$ for any ε, $0 \le \varepsilon \le 1$

$$|\int f_\varepsilon(|x-h|-s)P^i_{(n^{1/2})} * G^{n-i-1}_{n^{1/2}} * Q_{n^{1/2}}(dx)| \le c\varepsilon^{-2} n^{-3/2} \nu_1$$

$$+ c\bar{c}(V)[\sigma_1 + (n/(n-i-1))^{1/2} (1+n^{-1/2}(i \operatorname{tr} V)^{1/2} + |h|)](n/(n-i-1))^{1/2}\varepsilon^{-2}n^{-3/2}\nu_1$$

$$\le c(n,i)(1+\bar{c}(V))[1+(\operatorname{tr} V)^{1/2} + |h|]\varepsilon^{-2} \nu_1 . \tag{30}$$

Now if $(1+\sigma_1)(1+\bar{c}(V))\nu_1^{1/3} \le 1$ then combining (27) - (30) and (25) and putting $\varepsilon = \nu_1^{1/3}$ we obtain (26). If $(1+\sigma_1)(1+\bar{c}(V))\nu_1^{1/3} > 1$ the inequality (26) is obvious and the Lemma is proved. □

Note. The proof of (25) presented here belongs to V.V. Yurinskii.

Proof of Theorem 1. Without loss of generality we may assume $\mu = 0$. Suppose at first that $\sigma_1 = 1$. For $n < 4$ the theorem follows from (26). To prove it for $n \ge 4$ we shall use induction on n. Let $\varepsilon > 0$ be arbitrary and let f_ε be a function with properties (12). By Lemmas 1,2 we have

$$\Delta_n(h) = \sup_{r\ge 0} |P_n(S_r(h)) - G(S_r(h))| \le \sup_{s>0} |\int f_\varepsilon(|x-h|-s)(P_n - G)dx)| + c\bar{c}(V)(\sigma_1+|h|)\varepsilon . \tag{31}$$

Denoting $P_{(n)}$ the distribution of $n^{-1/2}X_1$,

$$G_t(\cdot) = G(t\cdot),\ t_i = (n/(n-i-1))^{1/2},\ i = \overline{0,n-2}$$

$$i_n = [n/2],\ H_i = P^i_{(n)} - G^i_{n^{1/2}},\ i = \overline{1,n}$$

$$V_i = P^i_{(n)} * G_{t_i},\ i = \overline{0,i_n},\quad V_i = H_i * G_{t_i},\ i = \overline{i_n+1,n-2}$$

$$V_{n-1} = H_{n-1}$$

and observing that

$$G^{n-i-1}_{n^{1/2}} = G_{t_i},\ G^i_{n^{1/2}} * G_{t_i} = G_{t_0}$$

we may write (cf. Ch. I, §2, (24) and (40))

$$P_n - G = P^n_{(n)} - G^n_{n^{1/2}}$$

$$= \left(\sum_{i=0}^{n-1} P^i_{(n)} * G^{n-i-1}_{n^{1/2}}\right) * H_1$$

$$= \left(\sum_{i=1}^{i_n} P^i_{(n)} * G_{t_i} + \sum_{i=i_n+1}^{n-2} H_i * G_{t_i} + H_{n-1} + (n-i_n)G_{t_0}\right) * H_1$$

$$= \left\{\sum_{i=1}^{n-1} V_i + (n - i_n)V_o\right\} * H_1 \; . \tag{32}$$

Applying Lemma 4 with $\alpha = n^{\frac{1}{2}}$, $\beta = t_i$, $\gamma = n^{\frac{1}{2}}$ we obtain for all $i = \overline{0, i_n}$, $0 < \varepsilon \leq 1$

$$|\int f_\varepsilon(|x - h| - s)V_i * H_1(dx)| \leq c\varepsilon^{-2}\nu n^{-3/2}$$

$$+ c\bar{c}(V)[\sigma_1 + t_i(1 + n^{-\frac{1}{2}}(i \text{ tr } V)^{\frac{1}{2}} + |h|)]t_i\varepsilon^{-2}\nu n^{-3/2}$$

$$\leq c\bar{c}(V)((\text{tr } V)^{\frac{1}{2}} + |h|)\varepsilon^{-2}\nu n^{-3/2} \; . \tag{33}$$

For $i = \overline{i_n+1, n-1}$ putting $G_{t_{n-1}} = \delta_o$ (the distribution concentrated at zero in H) we have as in (21)

$$|\int f_\varepsilon(|x - a| - s)V_i * H_1(dx)| \leq \sum_1^5 J_r \tag{34}$$

where

$$J_1 = \frac{1}{6}\iiint_{z_1 \in S_1} |\bar{g}_\varepsilon'''(\theta)| \; |H_i|(dx)G_{t_i}(dy)|H_1|(dz), \; |\theta| \leq 1$$

$$J_2 = \iiint_{z \in S_1^c} \bar{g}_\varepsilon(1) \; |H_i|(dx)G_{t_i}(dy)|H_1|(dz)$$

$$J_{3+j} = \iiint_{z \in S_1^c} |\bar{g}_\varepsilon^{(j)}(0)| \; |H_i|(dx)G_{t_i}(dy)|H_1|(dz), \; j = 0,1,2,$$

and $\bar{g}_\varepsilon(t)$ is defined by (18).

Observe now that (19) may be written as

$$\bar{g}_\varepsilon^{(j)}(t) = \bar{g}_\varepsilon^{(j)}(t)\chi_{S_{s,\varepsilon}(h-y-tz)}(x), \; j = 1,2,\ldots, \tag{35}$$

and the same argument as in (23) gives

$$J_1 \leq c\varepsilon^{-3}\int_{S_1} |z|^3 \; |H_1|(dz)\iint_{S_{s,\varepsilon}(h-y-tz)}(x) \; |H_i|(dx)G_{t_i}(dy). \tag{36}$$

Furthermore putting $g = h-y-\theta z$ we have

$$|H_i|(S_{s,\varepsilon}(g)) \leq (P^i_{(n)} + G^i_{n^{\frac{1}{2}}})(S_{s,\varepsilon}(g))$$

$$= H_i(S_{s,\varepsilon}(g)) + 2G^i_{n^{\frac{1}{2}}}(S_{s,\varepsilon}(g))$$

$$\leq 2 \sup_{s>0} |H_i(S_s(g))| + 2G^i_{n^{\frac{1}{2}}}(S_{s,\varepsilon}(g)) \; . \tag{37}$$

Moreover since for any Borel set B

$$H_i(B) = (P_i - G)(s_i B)$$

$$G^i_{n^{1/2}} = G_{s_i},\ G_{s_i}(B) = G(s_i B)$$

where $s_i = (n/i)^{1/2}$, by (37), Lemma 1 and the inductive hypothesis for all $z : |z| \le 1$, $i = \overline{i_n+1,\ n-1}$

$$|H_i|(S_{s,\varepsilon}(g)) \le 2 \sup_{s>0} |(P_i - G)(S_s(s_i g))| + 2G(S_{s_i s, s_i \varepsilon}(s_i g))$$

$$\le c\bar{c}\bar{c}(V)((\mathrm{tr}\ V)^{1/2} + s_i|g|)\nu^{1/3} i^{-1/6} + c\bar{c}(V)(\sigma_1 + s_i|g|)s_i\varepsilon$$

$$\le c\bar{c}(V)[(\mathrm{tr}\ V)^{1/2} + |h| + |y|](\bar{c}\nu^{1/3} n^{-1/6} + \varepsilon). \tag{38}$$

Note also that

$$\int |y| G_{t_i}(dy) \le ((n-i-1)/n)^{1/2} (\int |y|^2 G(dy))^{1/2}$$

$$= ((n-i-1)/n)^{1/2} (\mathrm{tr}\ V)^{1/2},\ i \le n-1\ . \tag{39}$$

Relations (36) - (39) imply

$$J_1 \le c\varepsilon^{-3} n^{-3/2} \nu \bar{c}(V)[(\mathrm{tr}\ V)^{1/2} + |h| + ((n-i-1)/n)^{1/2} (\mathrm{tr}\ V)^{1/2}](\bar{c}\nu^{1/3} n^{-1/6} + \varepsilon)$$

$$\le c\bar{c}(V)((\mathrm{tr}\ V)^{1/2} + |h|)(\bar{c}\varepsilon^{-3} \nu^{4/3} n^{-5/3} + \varepsilon^{-2} \nu n^{-3/2}).$$

To estimate $J_2 - J_5$ we use (12), Lemma 3 and the inequality

$$\int_{S_1^c} |z|^j |H_1|(dz) \le n^{-3/2}\nu,\ j = 0,1,2,$$

and obtain for all $\varepsilon : 0 < \varepsilon \le 1$

$$\sum_2^5 J_r \le c\varepsilon^{-2} \nu n^{-3/2}\ .$$

Thus for all $i = \overline{i_n+1,\ n-1}$, $0 < \varepsilon \le 1$

$$|\int f_\varepsilon(|x-h| - s) V_i * H_1(dx)| \le c\bar{c}(V)((\mathrm{tr}\,V)^{1/2} + |h|)(\bar{c}\varepsilon^{-3}\nu^{4/3} n^{-5/3} + \varepsilon^{-2} n^{-3/2}\nu)\ . \tag{40}$$

From (31) - (33) and (40) we deduce

$$\Delta_n(h) \le c\bar{c}(V)((\mathrm{tr}\ V)^{1/2} + |h|)[\varepsilon^{-2}\nu n^{-1/2} + \bar{c}\varepsilon^{-3}\nu^{4/3} n^{-2/3} + \varepsilon] \tag{41}$$

for all $\varepsilon : 0 < \varepsilon \le 1$.

We may assume that $\bar{c}$ is so large that $\bar{c} \ge 1$ and

$$c(\bar{c}^{-3/2} + 2\bar{c}^{-3/4}) \le 1$$

(here c is from (41). Then if

$$(\bar{c})^{1/4} \nu^{1/3} n^{-1/6} \le 1 \tag{42}$$

we may take ε in (41) to be equal to the left side of (42) and thus obtain

$$\Delta_n(h) \le \bar{\bar{c}}(V)((\mathrm{tr}\ V)^{\frac{1}{2}} + |h|)\nu^{1/3}n^{-1/6}[c(\bar{c}^{-3/2} + 2\bar{c}^{-3/4})]$$

$$\le \bar{\bar{c}}(V)((\mathrm{tr}\ V)^{\frac{1}{2}} + |h|)\nu^{1/3}n^{-1/6} .$$

If (42) is not satisfied the conclusion of the Theorem is obvious.

Thus we have proved the Theorem in the case $\sigma_1 = 1$. If $\sigma_1 \ne 1$ consider the sequence $X_j' = \sigma_1^{-1}X_j$, $j = 1,2,\ldots$. Marking the quantities corresponding to $\{X_j'\}$ with prime we have

$$\Delta_n'(h) = \Delta_n(h\sigma_1) \qquad V' = \sigma_1^{-2}V \text{ (hence } \sigma_1' = 1)$$

$$\bar{c}(V') = \sigma_1^2\bar{c}(V) \qquad \nu' = \sigma_1^{-3}\nu .$$

These formulas imply easily that the general case follows from the case $\sigma_1 = 1$. □

§ 2. A better speed estimate for balls with a fixed centre

In this section the speed of convergence in the Central Limit Theorem in Hilbert space will be improved to $n^{-\frac{1}{4}}$. However, this improved estimate is worse than that of Theorem 1, §1, in some other respects. The difference of the two estimates reflects the limits of the present progress in this area.

Theorem 1. Under the conditions and in the notations of Theorem 1, §1 if

$$E|X_1|^3 = \beta < \infty$$

then

$$\Delta_n \le c(S,\beta)(1 + |h|)^{3/2}n^{-\frac{1}{4}}, \quad n = 1,2,\ldots \tag{1}$$

where $S = (\sigma_1, \sigma_2, \ldots)$.

In what follows we use the notations of Theorem 1, §1 and assume that its conditions are satisfied. We may also suppose that $\mu = EX_1 = 0$.

Lemma 1. If Y is a random variable with distribution G then for the characteristic function $g(t,h,V)$ of $|Y + h|^2$ we have

$$g(t,h,V) = \int \exp\{it|x + h|^2\}G(dx)$$

$$= \exp\{it(R_t h,h)\}g(t,0,V) ,$$

where

$$R_t = (I - 2itV)^{-1}.$$

Moreover

$$g(t,0,V) = \prod_{j=1}^{\infty} (1 - 2it\sigma_j^2)^{-\frac{1}{2}}. \tag{2}$$

Proof. Let $\{e_1, e_2 \ldots\}$ be a basis in H constituted by the eigenvectors e_i corresponding to the eigenvalues σ_j^2 of V. The real random variables $Y_j = (Y, e_j)$ are independent normal $(0, \sigma_j)$. Denoting $h_j = (h, e_j)$ we have

$$\begin{aligned} g(t,h,V) &= E \exp\{it|Y+h|^2\} \\ &= E \exp\{it \sum_{j=1}^{\infty} (Y_j + h_j)^2\} \\ &= \prod_{j=1}^{\infty} E \exp\{it(Y_j + h_j)^2\} \\ &= \prod_{j=1}^{\infty} (2\pi)^{-\frac{1}{2}} \sigma_j^{-1} \int e^{it(x+h_j)^2 - x^2/2\sigma_j^2} dx \\ &= \prod_{j=1}^{\infty} \exp\{ith_j^2/(1 - 2it\sigma_j^2)\} (2\pi)^{-\frac{1}{2}} \sigma_j^{-1} \int e^{itx^2 - x^2/2\sigma_j^2} dx \end{aligned} \tag{3}$$

thus

$$g(t,h,V) = \exp\{it(R_t h, h)\} g(t,0,V).$$

Finally (2) follows from (3) and the well known formula for the characteristic function of the square of a normal $(0,\sigma)$ random variable

$$(2\pi)^{-\frac{1}{2}} \sigma^{-1} \int e^{itx^2 - x^2/2\sigma^2} dx = (1 - 2it\sigma^2)^{-\frac{1}{2}}. \quad \square$$

Denote by $F(k,\sigma)$ the class of all covariance operators such that $\sigma_k \geq \sigma$ (σ_k^2 being the k-th eigenvalue of V in the decreasing order). Obviously $F(k,\sigma) \supset F(k',\sigma')$ if $k' \geq k$, $\sigma \geq \sigma'$.

Corollary 1. In the notations of Lemma 1 for any integer k

$$|g(t,h,V)| \leq (1 + 4\sigma_k t^2)^{-k/4}.$$

This inequality implies in particular that if Y is a normal $(0,V)$ random variable with $V \in F(3,\sigma)$ then the distribution of $|Y+h|^2$ has a density which is bounded by some constant $c(\sigma)$.

Corollary 2. If X,Y are independent random variables with values in H and Y is normal (0,V) then

$$|E\exp\{it|h+X+Y|^2\}| \le |g(t,0,V)| .$$

Indeed

$$|E\exp\{it|h+X+Y|^2\}| = |E\, g(t,h+X,V)|$$
$$\le |g(t,0,V)| .$$

Lemma 2. The function

$$\hat{g}(\lambda) = g(t,h_1+\lambda h_2,V),\ \lambda \in R',\ h_1,\ h_2 \in H$$

is infinitely differentiable with respect to λ and

$$\hat{g}'(\lambda) = t(\gamma+\lambda\delta)\hat{g}(\lambda)$$

$$\hat{g}''(\lambda) = [t^2(\gamma+\lambda\delta)^2 + t\delta]\hat{g}(\lambda)$$

$$\hat{g}'''(\lambda) = [t^3(\gamma+\lambda\delta)^3 + 3t^2(\gamma+\lambda\delta)\delta]\hat{g}(\lambda)$$

$$\hat{g}^{(iv)}(\lambda) = [t^4(\gamma+\lambda\delta)^4 + 6t^3(\gamma+\lambda\delta)^2\delta + 3t^2\delta^2]\hat{g}(\lambda) \tag{4}$$

where

$$\gamma = 2i(R_t h_1,h_2)\ ,\quad \delta = 2i(R_t h_2,h_2) .$$

Moreover

$$|\gamma| \le 2|h_1|\ |h_2|\ ,\quad |\delta| \le 2|h_2|^2$$
$$|\hat{g}(\lambda)| \le |g(t,0,V)| . \tag{5}$$

Proof. Lemma 1 implies

$$\hat{g}(\lambda) = \exp\{it(R_t h_1,h_1) + \lambda t\alpha + (1/2)\lambda^2 t\beta\}g(t,0,V) .$$

Differentiating this equation we obtain (4). Inequalities (5) also follow immediately from formulas of Lemma 1. □

Lemma 3. For any α, $0<\alpha<1$, there exists a bounded (two-valued) real random variable $\xi=\xi(\alpha)$ such that

$$E\xi = 0,\ E\xi^2 = \alpha,\ E\xi^3 = 1.$$

Proof. The indicated properties are enjoyed by a random variable ξ which takes values

$$(1 \pm (1+4\alpha^3)^{1/2})/2\alpha$$

with probabilities

$$1/2 \pm 1/2(1+4\alpha^3)^{\frac{1}{2}} . \quad \square$$

Lemma 4. Let X be a random variable with values in H such that $EX=0$, $E|X|^3<\infty$. Denote V the covariance operator of X. Let Y be a normal $(0,(1-\alpha)V)$, $0<\alpha<1$, random variable and let $\xi=\xi(\alpha)$ be a real random variable with properties described in Lemma 3. Suppose that X,Y,ξ are independent. Then $X' = \xi X+Y$ has mean zero, covariance operator V and for any $h_1,h_2,h_3 \in H$

$$E(X,h_1)(X,h_2)(X,h_3) = E(X',h_1)(X',h_2)(X',h_3) \tag{6}$$

(in other words X and X' have the same moments of the first three orders). Moreover

$$E|X'|^m \le c(m,\alpha)E|X|^m, \; m \ge 2 . \tag{7}$$

Proof. Since ξ,X,Y are independent, $E\xi=0$, $EX=EY=0$ in H, $E\xi^2=\alpha$ and the covariance operators of X,Y are V, $(1-\alpha)V$ respectively, we have for any $h_1,h_2 \in H$

$$\begin{aligned} E(X',h_1) &= E\xi E(X,h_1)+E(Y,h_1) \\ &= 0 , \\ E(X',h_1)(X',h_2) &= E\xi^2 E(X,h_1)(X,h_2)+E(Y,h_1)(Y,h_2) \\ &= (Vh_1,h_2), \end{aligned}$$

i.e. $EX' = 0$ and the covariance operator of X' is V. Equation (6) is proved in a similar way: one has only to observe that the symmetry of the distribution of Y implies for any $h_1,h_2,h_3 \in H$

$$E(Y,h_1)(Y,h_2)(Y,h_3) = 0.$$

In order to prove (7) we shall show first that for any positive integer m

$$E|Y|^m \le c(m)(E|Y|^2)^{m/2} . \tag{8}$$

Indeed let $v_1 \ge v_2 \ge \dots$ be the eigenvalues of the covariance operator of Y and $e_1,e_2,\dots$ be the corresponding orthomormal eigenvectors. Denoting $Y_j = (Y,e_j)$ we have

$$E|Y|^m = E\left(\sum_{j=1}^{\infty} Y_j^2\right)^{m/2} \tag{9}$$

$$(E|Y|^2)^{m/2} = \left(\sum_{j=1}^{\infty} v_j\right)^{m/2} . \tag{10}$$

If m is even (8) is easily obtained by comparing (9) and (10) and observing that $E|Y_j|^{2\ell} = v_j^{\ell} c(\ell)$, $\ell > 0$. If m is odd we use the moment inequality

$$E|Y|^m \le (E|Y|^{m+1})^{m/(m+1)}$$

and apply (8) for even m.

To prove (7) we observe now that since

$$(a+b)^m \le 2^{m-1}(a^m + b^m), \quad a,b \ge 0,$$

we have

$$E|X'|^m \le 2^{m-1}(E|\xi|^m E|X|^m + E|Y|^m)$$

and, moreover by (8) and a standard moment inequality

$$\begin{aligned} E|Y|^m &\le c(m)(E|Y|^2)^{m/2} \\ &= c(m)((1-\alpha)\operatorname{tr} V)^{m/2} \\ &= c(m)(1-\alpha)^{m/2}(E|X|^2)^{m/2} \\ &\le c(m)(1-\alpha)^{m/2} E|X|^m . \end{aligned}$$

Together with the obvious inequality $E|\xi|^m \le c(m,\alpha)$ this proves (7). □

Lemma 5. Let X_i, Y_i, $j = 1,\dots,n$ be H valued and ξ_j, $j = 1,\dots,n$ be real random variables. We assume that X_j, Y_j, ξ_j, $j = 1,\dots,n$ are independent. Assume also that X_j, $j = 1,\dots,n$ are identically distributed, $|X_j| \le B$, $EX_j = 0$ and that the covariance operator V of X_j belongs to $F(3,\sigma)$. Finally we suppose that Y_j are normal $(0,(1-\alpha)V)$, $0 < \alpha < 1$ and $\xi_j = \xi_j(\alpha)$ satisfy the conditions of Lemma 3. Put $X_j' = \xi_j X_j + Y_j$ and let P_n (resp. P_n') be the distribution of $n^{-\frac{1}{2}}\sum_1^n X_j$ (resp. $n^{-\frac{1}{2}}\sum_1^n X_j'$). Then for any $h \in H$

$$\begin{aligned} \delta_n &= \sup_{r \ge 0} |P_n(S_r(h)) - P_n'(S_r(h)| \\ &\le c(\alpha,\sigma,B)(1+|h|)n^{-\frac{1}{4}} \end{aligned} \tag{11}$$

Proof. Denote by $P_{(n)}$ (resp. $P'_{(n)}, Q_{(n)}, G_{(n)}$) the distribution of $n^{-\frac{1}{2}}X_j$ (resp. $n^{-\frac{1}{2}}X_j'$, $n^{-\frac{1}{2}}\xi_j X_j, n^{-\frac{1}{2}}Y_j$). Then

$$\begin{aligned} P_n - P_n' &= P_{(n)}^n - (P'_{(n)})^n \\ &= (P_{(n)} - P'_{(n)}) * \sum_{j=0}^{n-1} P_{(n)}^j * Q_{(n)}^{n-j-1} * G_{(n)}^{n-j-1} . \end{aligned}$$

Note that since $G_{(n)}^{n-j-1}$ is normal $(0,V_{j,\alpha})$ distribution, where $V_{j,\alpha} = (1-(j+1)/n)\times$ $\times\ (1-\alpha)V$, we have by Lemma 1

$$\int \exp\{it|h+x+z|^2\} G_{(n)}^{n-j-1}(dz) = g(t,h+x,V_{j,\alpha}),\ x \in H .$$

Hence

$$\begin{aligned}\delta_n(t) &= \int \exp\{it|h+x|^2\}(P_n - P_n')(dx) \\ &= \sum_{j=0}^{n-1} \iint g(t,h+x+y,V_{j,\alpha})(P_{(n)} - P_{(n)}')(dx)(P_{(n)}^j * Q_{(n)}^j * Q_{(n)}^{n-j-1})(dy) \\ &= \sum_{j=0}^{n-1} I_j .\end{aligned}$$

Denote for brevity

$$\hat{g}_{1,j}(\lambda) = g(t,h+\lambda x+y,V_{j,\alpha}),\ \lambda \in R$$

and observe that by (4) and Lemma 4

$$\int \hat{g}_{1,j}^{(k)}(0)(P_{(n)} - P_{(n)}')(dx) = 0,\ k = 0,1,2,3.$$

Thus expanding $\hat{g}_{1,j}(\lambda)$ by Taylor's formula up to the term of the fourth order we obtain

$$|I_j| \le (1/4!)\iint \sup_{0\le\lambda\le 1} |\hat{g}_{1,j}^{(iv)}(\lambda)|(P_{(n)} + P_{(n)}')(dx)(P_{(n)}^j * Q_{(n)}^{n-j-1})(dy). \tag{12}$$

Furthermore by Lemma 2

$$\begin{aligned}\sup_{0\le\lambda\le 1} |\hat{g}_{1,j}^{(iv)}(\lambda)| \le c\{t^4[(|h|^4+|y|^4)|x|^4 + |x|^8] \\ + |t|^3[|h|^2 + |y|^2)|x|^2 + |x|^4]|x|^2 + t^2|x|^4]|g(t,0,V_{j,\alpha})| .\end{aligned} \tag{13}$$

The assumption $|X_j| \le B$ and Lemma 4 imply also

$$\begin{aligned}\int |x|^m P_{(n)}(dx) &= E|n^{-\frac{1}{2}}X_1|^m \\ &\le B^m n^{-m/2} , \\ \int |x|^m P_{(n)}'(dx) &= E|n^{-\frac{1}{2}}X_1'|^m \\ &\le c(m,\alpha)B^m n^{-m/2} .\end{aligned} \tag{14}$$

Finally by the properties of X_j and ξ_j we have

$$\begin{aligned}\int |y|^2 P_{(n)}^j * Q_{(n)}^{n-j-1}(dy) &= E\Big|n^{-\frac{1}{2}}\Big(\sum_{r=1}^{j} X_r + \sum_{r=j+2}^{n} \xi_r X_r\Big)\Big|^2 \\ &= n^{-1}E\Big(\sum_{r=1}^{j} |X_r|^2 + \sum_{r=j+2}^{n} \xi_r^2 |X_r|^2\Big) \le B^2\end{aligned} \tag{15}$$

and similarly

$$\int |y|^4 P^j_{(n)} * Q^{n-j-1}_{(n)}(dy) = E\left|n^{-\frac{1}{2}}\left(\sum_{r=1}^{j} X_r + \sum_{r=j+2}^{n} \xi_r X_r\right)\right|^4$$

$$\leq B^4 c(\alpha) \tag{16}$$

(the last inequality follows from the implication: if $E(X_{r_1}, X_{r_2})(X_{r_3}, X_{r_4}) \neq 0$ then $r_1 = r_2 = r_3 = r_4$ of $r_1 = r_2$, $r_3 = r_4$ or $r_1 = r_3$, $r_2 = r_4$ or $r_1 = r_4$, $r_2 = r_3$). From (12) - (16) we deduce now

$$I_j \leq c(\alpha, B) n^{-2} (1 + |h|)^4 t^2 (1 + t^2) |g(t, 0, V_{j,\alpha})| ,$$

so that

$$|\delta_n(t)| \leq c(\alpha, B) n^{-2} (1 + |h|^4) t^2 (1 + t^2) \sum_{j=0}^{n-1} |g(t, 0, V_{j,\alpha})| . \tag{17}$$

Note now that the distribution of $|h + n^{-\frac{1}{2}} \sum_1^n X_j|^2$ has a bounded density $p(u)$. Indeed, denoting $X = n^{-\frac{1}{2}} \sum_1^n \xi_j X_j$, $Y = n^{-\frac{1}{2}} \sum_1^n Y_j$ we have

$$P(|h + X + Y|^2 < u) = \int P(|h + x + Y|^2 < u) P_X(dx)$$

and by Corollary 1 the distribution of $|h + x + Y|^2$ has a density $q(x,u)$ which is bounded by a constant $c(\sigma)$. Hence $p(u)$ exists and

$$p(u) = \int q(x,u) P_X(dx)$$
$$\leq c(\sigma) .$$

Applying Esseen inequality [15] and using (17) we may thus write

$$\delta_n \leq c \int_0^T |\Delta_n(t)| t^{-1} dt + c(\sigma) T^{-1} \tag{18}$$

$$\leq c(\alpha, \sigma, B) [n^{-2} (1 + |h|)^4 \int_0^T t(1 + t^2) \sum_{j=0}^{n-1} |g(t, 0, ((j+1)/n)(1 - \alpha)V)| dt + T^{-1}].$$

Furthermore it follows from (2) that

$$|g(t, 0, u(1 - \alpha)V| = |g(ut, 0, (1 - \alpha)V)|$$

as a function of $u \geq 0$ is nonincreasing and hence

$$n^{-1} \sum_{j=0}^{n-1} |g(t, 0, ((j+1)/n)(1 - \alpha)V)| \leq \int_0^1 |g(ut, 0, (1 - \alpha)V)| du$$

$$\leq t^{-1} \int_0^\infty |g(v, 0, (1 - \alpha)V)| dv = c(\alpha, \sigma) t^{-1}. \tag{19}$$

Together (18) and (19) imply for any $T \geq 1$

$$\delta_n \leq c(\alpha,\sigma,B)[n^{-1}(1+|h|)^4 T^3 + T^{-1}] . \tag{20}$$

If

$$(1+|h|)n^{-\frac{1}{4}} < 1 \tag{21}$$

we can take in (20) $T = (1+|h|)^{-1} n^{-\frac{1}{4}}$ and (11) follows. If (21) is not satisfied the validity of (11) is obvious. □

Let X be a random variable with values in H such that $EX = 0$, $\beta = E|X|^3 < \infty$. Denote by $\hat{X}$ the truncation of X at a level $R > 0$, i.e.

$$\hat{X} = \begin{cases} X & \text{if } |X| < R \\ 0 & \text{if } |X| \geq R . \end{cases} \tag{22}$$

Putting $P = P_X$, $\hat{P} = P_{\hat{X}}$ we have for any Borel set A

$$\hat{P}(A) = P(A \cap S_R) + P(S_R^c)\chi_A(0) .$$

If $P(S_{R_1}) > 0$, $R_1 > 0$, define also a measure P_1 by

$$P_1(A) = P(A \cap S_{R_1})/P(S_{R_1}) .$$

<u>Lemma 6</u>. If $R \geq R_1$ and $P(S_{R_1}) \geq 1/2$ then

$$\hat{P} \geq (1/2)P_1$$

so that

$$\hat{P} = (1/2)P_1 + (1/2)P_2 \tag{23}$$

where P_2 is a probability measure concentrated at S_R.

Furthermore for any positive integer k if $R_1 \geq c(k,\beta)$ then

$$\sigma_{1j}^2 \geq (1/2)\sigma_j^2 \ , \ j = 1,\dots,k \tag{24}$$

where $\sigma_1^2 \geq \sigma_2^2$ ... (resp. $\sigma_{1j}^2 \geq \sigma_{2j}^2 \geq \dots$) are the eigenvalues of the covariance operator of P (resp. P_1). Similarly if $R_1 \geq c'(k,\beta)$ then

$$\hat{\sigma}_{1j}^2 \geq (3/4)\sigma_j^2 \ , \ j = 1,\dots,k \tag{24'}$$

where $\hat{\sigma}_{1j}^2 \geq \hat{\sigma}_{2j}^2 \geq \dots$ are the eigenvalues of the covariance operator of the measure

$$\hat{P}_1(A) = P(A \cap S_{R_1}) + P(S_{R_1}^c)\chi_A(0) .$$

Moreover

$$\left|\int x\hat{P}(dx)\right| \leq \beta R^{-2}$$

and if $P(S_{R_1}) \geq 1/2$ then

$$\int |x|^3 P_2(dx) \leq 2\beta \ .$$

<u>Remark</u>. Clearly there exist $c_1(k,\beta)$ such that if

$$R_1 \geq c_1(k,\beta) \tag{25}$$

then $R_1 \geq c(k,\beta)$, $R_1 \geq c'(k,\beta)$ and $P(S_{R_1}) \geq 1/2$ simultaneously.

<u>Proof</u>. If $R \geq R_1$, $P(S_{R_1}) \geq 1/2$ we have for any Borel set A

$$\begin{aligned}\hat{P}(A) &\geq P(A \cap S_R)\\ &\geq P(A \cap S_{R_1})\\ &\geq (1/2)P(A \cap S_{R_1})/P(S_{R_1})\\ &= (1/2)P_1(A) \ .\end{aligned}$$

Furthermore the theory of perturbation of linear operators in Hilbert space implies that to prove (24) it is enough to show that for any $\varepsilon > 0$ we have

$$\sup_{h\in S_1} |(V - V_1)h| < \varepsilon \text{ if } R_1 \geq c(\varepsilon,\beta) \tag{26}$$

where V_1 is the covariance operator of P_1. Denoting

$$\begin{aligned}p_1 &= P(S_{R_1})\\ \mu_1 &= \int x P_1(dx)\\ &= p_1^{-1}\int_{S_{R_1}} xP(dx)\\ &= -\ p_1^{-1}\int_{S_{R_1}^c} xP(dx)\end{aligned}$$

we may write

$$\begin{aligned}(Vh_1,h_2) &= \int (h_1,x)(h_2,x)P(dx)\\ (V_1h_1,h_2) &= \int (h_1,x)(h_2,x)P_1(dx) - (h_1,\mu_1)(h_2,\mu_2)\\ &= p_1^{-1}\int_{S_{R_1}} (h_1,x)(h_2,x)P(dx)\\ &\quad - p_1^{-2}\int_{S_{R_1}^c}(h_1,x)P(dx)\int_{S_{R_1}^c}(h_2,x)P(dx) \ .\end{aligned}$$

Consequently

$$|((V-V_1)h_1,h_2)| \le [(1-p_1^{-1})\int_{S_{R_1}}|x|^2P(dx)+\int_{S_{R_1}^c}|x|^2P(dx)$$

$$+\,p_1^{-2}(\int_{S_{R_1}^c}|x|P(dx))^2]|h_1|\;|h_2|\ ,$$

and (26) follows. Relation (24') is proved in a similar way. Finally

$$\begin{aligned}|\int_{S_R} xP(dx)| &= |\int_{S_R^c} xP(dx)| \\ &\le R^{-2}\int_{S_R^c}|x|^3P(dx) \\ &\le \beta R^{-2}\ ,\end{aligned}$$

and if $p_1 \ge 1/2$, then $P_2 \le 2P$ and

$$\int|x|^3P_2(dx) \le 2\beta\ .\ \square$$

Preserving the notations of Lemma 6 denote by X' a P_1-distributed random variable and let $\xi=\xi(1/2)$ be a real random variable described in Lemma 3 and Y be a $(0,(1/2)V_1)$ normal random variable, where V_1 is the covariance operator of P_1. The random variables X',ξ,Y are supposed to be independent. Let $\tilde P_1$ be the distribution of

$$EX' + \xi(X'-EX') + Y$$

and put (assuming that $R\ge R_1$, $P(S_{R_1})\ge 1/2$)

$$\tilde P = (1/2)\tilde P_1 + (1/2)P_2\ . \tag{27}$$

<u>Lemma 7</u>. For any $r\ge 0$

$$\int|x|^{3+r}\tilde P(dx) \le c(r)\beta R^r\ .$$

<u>Proof</u>. By Lemma 4 for any $s\ge 2$

$$\begin{aligned}\int|x|^s\tilde P_1(dx) &= E|EX' + \xi(X'-EX') + Y|^s \\ &\le c(s)(|EX'|^s + E|\xi(X'-EX') + Y|^s) \\ &\le c(s)(E|X'|^s + E|X'-EX'|^s) \\ &\le c(s)E|X'|^s \\ &= c(s)\int|x|^sP_1(dx)\ .\end{aligned}$$

Hence, since $\tilde{P} = (\tilde{P}_1 + P_2)/2$, $\hat{P} = (P_1 + P_2)/2$,

$$\begin{aligned}\int |x|^{3+r}\tilde{P}(dx) &\le c(r)\int |x|^{3+r}\hat{P}(dx) \\ &\le c(r)R^r \int |x|^r \hat{P}(dx) \\ &\le c(r)\beta R^r . \quad \square\end{aligned}$$

Lemma 8. Let P be a probability measure with the same properties as in Lemma 6 and with covariance matrix $V \in F(k,\sigma)$, $k \ge 3$. Suppose that $R_1 \le R$ and that R_1 satisfies (25). Let $\hat{P}$, $\tilde{P}$ be probability measures defined by (23), (27). Furthermore let $\hat{X}_j$ (resp. $\tilde{X}_j$) be independent random variables with

$$P_{\hat{X}_j} = \hat{P} \text{ (resp. } P_{\tilde{X}_j} = \tilde{P})$$

and denote

$$\hat{S}_n = n^{-\frac{1}{2}} \sum_1^n \hat{X}_j \quad , \quad \hat{P}_n = P_{\hat{S}_n}$$

$$\tilde{S}_n = n^{-\frac{1}{2}} \sum_1^n \tilde{X}_j \quad , \quad \tilde{P}_n = P_{\tilde{S}_n} .$$

Then for any $h \in H$, $r \ge 0$

$$|\hat{P}_n(S_r(h)) - \tilde{P}_n(S_r(h))| \le c(\beta,\sigma)(1 + |h| + n^{\frac{1}{4}}R^{-2})n^{-\frac{1}{4}} \tag{28}$$

and

$$|E \exp\{it|h + \tilde{S}_n|^2\}| \le |g(t,0,(1/8)V_1)| + \exp\{-c'n\} \tag{29}$$

where g is the same as in Lemma 1 and $c' > 0$.

Proof. Denote

$$\mu_1 = \int x P_1(dx) = \int x \tilde{P}_1(dx) \quad , \quad \mu_2 = \int x P_2(dx)$$

and for any Borel set A and integer $j > 0$ let

$$\bar{P}_1(A) = P_1(A - \mu_1) \qquad \bar{\tilde{P}}_1(A) = \tilde{P}_1(A - \mu_1)$$

$$\bar{P}_2(A) = P_2(A - \mu_2)$$

$$(\bar{P}_1)_j(A) = (\bar{P}_1)^j(j^{\frac{1}{2}}A) \quad (\bar{\tilde{P}}_1)_j(A) = (\bar{\tilde{P}}_1)^j(j^{\frac{1}{2}}A) .$$

Then, since for any $\alpha > 0$, $f \in H$

$$\alpha(S_r(h) + f) = S_{\alpha r}(\alpha f + \alpha h)$$

we have

$$\hat{P}_n(S_r(h)) = P(\hat{S}_n \in S_r(h))$$

$$= (2^{-1}(P_1 + P_2))^n(n^{\frac{1}{2}}S_r(h))$$

$$= \sum_{j=0}^{n} \binom{n}{j} 2^{-n} P_1^j * P_2^{n-j}(n^{\frac{1}{2}}S_r(h))$$

$$= \sum_{j=0}^{n} \binom{n}{j} 2^{-n} \bar{P}_1^j * \bar{P}_2^{n-j}(n^{\frac{1}{2}}S_r(h) - j\mu_1 - (n-j)\mu_2)$$

$$= 2^{-n}\bar{P}_2(n^{\frac{1}{2}}S_r(h) - n\mu_2)$$

$$+ \sum_{j=1}^{n} \binom{n}{j} 2^{-n} \int (\bar{P}_1)_j (j^{-\frac{1}{2}}(n^{\frac{1}{2}}S_r(h) - j\mu_1 - (n-j)\mu_2 - x))\bar{P}_2^{n-j}(dx)$$

$$= 2^{-n}\bar{P}_2(n^{\frac{1}{2}}S_r(h) - n\mu_2)$$

$$+ \sum_{j=1}^{n} \binom{n}{j} 2^{-n} \int (\bar{P}_1)_j (S_{(n/j)^{\frac{1}{2}}r}((n/j)^{\frac{1}{2}}h - j^{\frac{1}{2}}\mu_1 - (n-j)j^{-\frac{1}{2}} - j^{-\frac{1}{2}}x)\bar{P}_2^{n-j}(dx). \quad (30)$$

Similarly

$$\tilde{P}_n(S_r(h)) = 2^{-n}\bar{P}_2(n^{\frac{1}{2}}S_r(h) - n\mu_2)$$

$$+ \sum_{j=1}^{n} \binom{n}{j} 2^{-n} \int (\bar{\tilde{P}}_1)_j (S_{(n/j)^{\frac{1}{2}}r}((n/j)^{\frac{1}{2}}h - j^{\frac{1}{2}}\mu_1 - (n-j)j^{-\frac{1}{2}} - j^{-\frac{1}{2}}x)\bar{P}_2^{n-j}(dx). \quad (31)$$

By Lemma 5 the absolute value of the difference of the integrands in (30) and (31) is not greater than

$$D_j(x) = c(\alpha,\beta)(1 + (n/j)^{\frac{1}{2}}|h| + |j^{\frac{1}{2}}\mu_1 + (n-j)j^{-\frac{1}{2}}\mu_2| + j^{-\frac{1}{2}}|x|)j^{-\frac{1}{4}} .$$

Furthermore by Lemma 6

$$|j^{\frac{1}{2}}\mu_1 + (n-j)j^{-\frac{1}{2}}\mu_2| = |j^{\frac{1}{2}}(\mu_1 + \mu_2) + j^{-\frac{1}{2}}(n-2j)\mu_2|$$

$$\leq 2j^{\frac{1}{2}}\beta R^{-2} + j^{-\frac{1}{2}}|n-2j|(2\beta)^{1/3}$$

and

$$\int |x| \bar{P}_2^{n-j}(dx) \leq (\int |x|^2 \bar{P}_2^{n-j}(dx))^{\frac{1}{2}}$$

$$= (\int |\sum_{r=1}^{n-j} x_r|^2 \bar{P}_2(dx_1) \ldots \bar{P}_2(dx_{n-j}))^{\frac{1}{2}}$$

$$= (n-j)^{\frac{1}{2}}(\int |x|^2 \bar{P}_2(dx))^{\frac{1}{2}}$$

$$\leq (n-j)^{\frac{1}{2}}(\int |x|^2 P_2(dx))^{\frac{1}{2}}$$

$$\leq (n-j)^{\frac{1}{2}}(2\beta)^{1/3} .$$

Consequently

$$\int D_j(x)\bar{P}_2^{n-j}(dx) \le c(\sigma,\beta)(1+(n/j)^{\frac{1}{2}}|h| + j^{\frac{1}{2}}R^{-2} + j^{-\frac{1}{2}}|n-2j| + j^{-\frac{1}{2}}(n-j)^{\frac{1}{2}})j^{-\frac{1}{4}} .$$

Thus we obtain

$$\begin{aligned}|\hat{P}_n(S_r(h))-\tilde{P}_n(S_r(h))| &\le c(\sigma,\beta)\sum_{j=1}^{n}\binom{n}{j}2^{-n}(j^{-\frac{1}{4}} + j^{-3/4}n^{\frac{1}{2}}|h| \\ &\quad + j^{\frac{1}{4}}R^{-2} + j^{-3/4}|n-2j| + j^{-3/4}(n-j)^{\frac{1}{2}}) \\ &\le c(\sigma,\beta)\big(E(1+\eta)^{-\frac{1}{4}} + n^{\frac{1}{2}}|h|E(1+\eta)^{-3/4} \\ &\quad + R^{-2}E\eta^{\frac{1}{4}} + E(1+\eta)^{-3/4}|n-2\eta| + E(1+\eta)^{-3/4}(n-\eta)^{\frac{1}{2}}\big) \qquad (32)\end{aligned}$$

where η is binomially distributed, i.e. distributed as the sum of n independent real random variables η_i such that $P(\eta_i = 0) = P(\eta_i = 1) = 1/2$. Note now that by Bernstein inequality for any $\gamma > 0$ we have

$$\begin{aligned}E(1+\eta)^{-\gamma} &\le E(1+\eta)^{-\gamma}\chi_{\eta>n/4} + P(\eta < n/4) \\ &\le (1+n/4)^{-\gamma} + e^{-3n/32} \\ &\le c(\gamma)n^{-\gamma} . \qquad (33)\end{aligned}$$

Moreover

$$\begin{aligned}E(1+\eta)^{-3/4}|n-2\eta| &\le (E(1+\eta)^{-3/2}E(n-2\eta)^2)^{\frac{1}{2}} \\ &= n^{\frac{1}{2}}(E(1+\eta)^{-3/2})^{\frac{1}{2}} \\ E(1+\eta)^{-3/4}(n-\eta)^{\frac{1}{2}} &\le (E(1+\eta)^{-3/2}E(n-\eta))^{\frac{1}{2}} \\ &= (n/2)^{\frac{1}{2}}(E(1+\eta)^{-3/2})^{\frac{1}{2}} \qquad (34)\end{aligned}$$

and, obviously

$$\begin{aligned}E\eta^{\frac{1}{4}} &\le (E\eta)^{\frac{1}{4}} \\ &= (n/2)^{\frac{1}{4}} . \qquad (35)\end{aligned}$$

Relations (32) - (35) imply (28).

To prove (29) we note that $\tilde{P} = (1/2)(Q * G_1 + P_2)$ where Q is a probability measure and G_1 is normal $(0,(1/2)V_1)$ (see (27) and the definition of $\tilde{P}_1$ above). Observing also that $G_1^j(n^{\frac{1}{2}}A)$ as a function of Borel sets A is the normal $(0,(j/2n)V_1)$ probability measure, we have (see Lemma 1)

$$E\exp\{it|h+\tilde{S}_n|^2\} = \sum_{j=0}^{n}\binom{n}{j}2^{-n}\int\exp\{it|h+n^{\frac12}x|^2\}Q^j * G_1^j * P_2^{n-j}(dx)$$

$$= \sum_{j=0}^{n}\binom{n}{j}2^{-n}\int\exp\{it|h+x+n^{-\frac12}y|^2\}G_1^j(n^{\frac12}dx)Q^j * P_2^{n-j}(dy)$$

$$= \sum_{j=0}^{n}\binom{n}{j}2^{-n}\int g(t,h+n^{-\frac12}y,(j/2n)V_1)Q^j * P_2^{n-j}(dy) \ . \tag{36}$$

Furthermore for $j > n/4$ the formulas of Lemma 1 imply

$$|g(t,h+n^{-\frac12}y,(j/2n)V_1)| \le |g(t,0,(j/2n)V_1)|$$

$$\le |g(t,0,(1/8)V_1)| \ . \tag{37}$$

Moreover applying e.g. Bernstein inequality we have

$$\sum_{j\le n/4}\binom{n}{j}2^{-n} \le e^{-3n/32} \ . \tag{38}$$

Relations (36) - (38) imply (29). □

<u>Proof of Theorem 2</u>. We may assume that $n^{\frac12} \ge R_1 = c_1(4,\beta)$ (see (25)) since otherwise (1) is obtained by a proper choice of $c(S,\beta)$.

Let $\hat{X}_j$ be the truncation of X_j at the level $R = n^{\frac12}$ (see (22)) and let $\hat{P}_n$ be the distribution of $\hat{S}_n = n^{-\frac12}\sum_1^n \hat{X}_j$. Denoting

$$A_j = \{\omega: X_j = \hat{X}_j\} \qquad A = \bigcap_{j=1}^{n} A_j$$

we have

$$|P_n(S_r(h)) - \hat{P}_n(S_r(h))| \le |P(\{S_n \in S_r(h)\} \cap A) + P(\{\hat{S}_n \in S_r(h)\} \cap A)| + P(A^c)$$

$$\le \sum_{j=1}^{n} P(A_j^c)$$

$$= nP(|X_1| \ge n^{\frac12})$$

$$\le \beta n^{-\frac12} \ . \tag{39}$$

Note that

$$E|\hat{X}_1| = |EX_1\chi_{A_1}|$$

$$= |EX_1\chi_{A_1^c}|$$

$$\le E|X_1|\chi_{A_1^c}$$

$$\le \beta n^{-1} \ . \tag{40}$$

For the covariance operator $\hat{V}$ of $\hat{X}_1$ we have

$$(\hat{V}f,f) = E(X_1\chi_{A_1},f)^2 - (EX_1\chi_{A_1},f)^2$$
$$\leq E(X_1,f)^2$$
$$\leq (Vf,f) . \tag{41}$$

Furthermore, since $E(X_1,f)\chi_{A_1} = -E(X_1,f)\chi_{A_1^c}$ we may write

$$(Vf,f) - (\hat{V}f,f) = E(X_1,f)^2\chi_{A_1^c} + (E(X_1,f)\chi_{A_1})^2$$
$$\leq 2E(X_1,f)^2\chi_{A_1^c} ,$$

and denoting $\{e_j\}$ an orthonormal basis in H we obtain

$$\mathrm{tr}(V-\hat{V}) \leq 2\sum_{j=1}^{\infty} E(X_1,e_j)^2\chi_{A_1^c}$$
$$= 2E|X_1|^2\chi_{A_1^c}$$
$$\leq 2\beta n^{-\frac{1}{2}} . \tag{42}$$

Let now $\hat{Y}$, Y' be two independent normal H-valued random variables with parameters $(n^{\frac{1}{2}}\hat{\mu},\hat{V})$, $(0,V-\hat{V})$ respectively, where $\hat{\mu} = E\hat{X}_1$. Obviously the random variable

$$Y = \hat{Y} - n^{\frac{1}{2}}\hat{\mu} + Y'$$

is normal $(0,V)$, i.e. $P_Y = G$. Furthermore, since

$$\{Y \in S_r(h)\} = \{\hat{Y} \in S_r(h + n^{\frac{1}{2}}\hat{\mu} - Y')\}$$

and since for any $h,h_1 \in H$, $r > 0$

$$S_r(h)\Delta S_r(h_1) \subset S_{r-|h-h_1|,2|h-h_1|}^{(h)} ,$$

applying Lemma 1, §1 we have

$$|P_Y(S_r(h)) - P_{\hat{Y}}(S_r(h))| = |P_{\hat{Y}}(S_r(h+n^{\frac{1}{2}}\hat{\mu} - Y')) - P_{\hat{Y}}(S_r(h))|$$
$$\leq P_{\hat{Y}}(S_r(h+n^{\frac{1}{2}}\hat{\mu} - Y')\Delta S_r(h))$$
$$\leq \int P_{\hat{Y}}(S_{r-n^{\frac{1}{2}}|\hat{\mu}|-|y|,2(n^{\frac{1}{2}}|\hat{\mu}|+|y|)}^{(h)})P_{Y'}(dy)$$
$$\leq 2c(\hat{V})(\hat{\sigma}_1 + |h|)(n^{\frac{1}{2}}|\hat{\mu}| + E|Y'|)$$

where

$$c(\hat{V}) = c\hat{\sigma}_1^{-1}\hat{\sigma}_2^{-1}\prod_{j=3}^{\infty}(1-\hat{\sigma}_j^2/2\hat{\sigma}_3^2)^{-\frac{1}{2}}$$

and $\hat{\sigma}_1^2 \ge \hat{\sigma}_2^2 \ge \ldots$ are the eigenvalues of $\hat{V}$. Since $\hat{\sigma}_j \le \sigma_j$, $j = 1,2,\ldots$ (this follows from the mini-max property of the eigenvalues), (24') implies

$$c(\hat{V}) \le c_1(S) = (4/3)\sigma_1^{-1}\sigma_2^{-1}\prod_{j=3}^{\infty}(1-2\sigma_j^2/3\sigma_j^2)^{-\frac{1}{2}}.$$

Moreover by (40)

$$|\hat{\mu}| \le \beta n^{-1}$$

and by (42)

$$\begin{aligned} E|Y'| &\le (E|Y'|^2)^{\frac{1}{2}} \\ &\le (\mathrm{tr}(V-\hat{V}))^{\frac{1}{2}} \\ &\le (2\beta)^{\frac{1}{2}}n^{-\frac{1}{4}}. \end{aligned}$$

Thus

$$|P_Y(S_r(h)) - P_{\hat{Y}}(S_r(h))| \le c(S,\beta)(1+|h|)n^{-\frac{1}{4}}. \tag{43}$$

Now (39) and (43) reduce the proof of the theorem to the estimation of

$$|\hat{P}_n(S_r(h)) - P_{\hat{Y}}(S_r(h))|.$$

Applying Lemma 8 we may write

$$\begin{aligned} |\hat{P}_n(S_r(h)) - P_{\hat{Y}}(S_r(h))| &\le |\tilde{P}_n(S_r(h)) - P_{\tilde{Y}}(S_r(h))| \\ &\quad + c(\beta,\sigma)(1+|h|)n^{-\frac{1}{4}}; \end{aligned} \tag{44}$$

here $\tilde{P}_n$ is the distribution of $n^{-\frac{1}{2}}\sum_1^n \tilde{X}_j$, where $\tilde{X}_j$ have the same distribution (27) constructed starting with $P = P_{X_j}$ and with $R_1 = c_1(4,\beta)$ (see (25)), $R = n^{\frac{1}{2}}$.

To estimate

$$D_n = \sup_{r\ge 0}|\tilde{P}_n(S_r(h)) - P_{\hat{Y}}(S_r(h))|$$

we use the same method as in the proof of Lemma 5. Let $\tilde{P}_{(n)}$ be the distribution of $n^{-\frac{1}{2}}\tilde{X}_1$ and $G_{(n)}$ be the normal $(n^{-\frac{1}{2}}\hat{\mu}, n^{-1}\hat{V})$ distribution, so that $\tilde{P}_n = \tilde{P}_{(n)}^n$, $P_{\hat{Y}} = \tilde{G}_{(n)}^n$. We have

$$\begin{aligned} \tilde{P}_n - P_{\hat{Y}} &= \tilde{P}_{(n)}^n - \tilde{G}_{(n)}^n \\ &= (\tilde{P}_{(n)} - \tilde{G}_{(n)}) * \sum_{j=0}^{n-1}\tilde{P}_{(n)}^j * \tilde{G}_{(n)}^{n-j-1}, \end{aligned}$$

and, since $\tilde{G}_{(n)}^{n-j-1}$ is normal $((n-j-1)n^{-\frac{1}{2}}\hat{\mu},\ (n-j-1)n^{-1}\hat{V})$

$$\int \exp\{it|h+x+z|^2\}\tilde{G}_n^{n-j-1}(dz) = g(t, h+x+(n-j-1)n^{-\frac{1}{2}}\hat{\mu},\ (n-j-1)n^{-1}\hat{V}).$$

Hence denoting

$$\hat{g}_{2,j}(\lambda) = g(t, h+\lambda x+y+(n-j-1)n^{-\frac{1}{2}}\hat{\mu},\ (n-j-1)n^{-1}\hat{V})$$

we may write

$$\begin{aligned} d_n(t) &= \int \exp\{it|h+x|^2\}(\tilde{P}_n - P_{\hat{Y}})(dx) \\ &= \iint \hat{g}_{2,j}(1)(\tilde{P}_{(n)} - \tilde{G}_{(n)})(dx)\tilde{P}_{(n)}^j(dy) \\ &= \sum_{j=0}^{n-1} J_j . \end{aligned} \tag{45}$$

Using (4) and observing that $\tilde{P}_{(n)}$ and $\tilde{G}_{(n)}$ have the same first and second moments (recall that $\tilde{X}_j$ and $\hat{X}_j$ have the same first, second and even third moments) we have

$$|J_j| \le (1/6)\int \sup_{0\le\lambda\le1} |\hat{g}_{2,j}'''(\lambda)|(\tilde{P}_{(n)} + \tilde{G}_{(n)})(dx)\tilde{P}_{(n)}^j(dy) . \tag{46}$$

Furthermore by Lemma 2

$$\begin{aligned} \sup_{0\le\lambda\le1} |\hat{g}_{2,j}'''(\lambda)| \le c[|t|^3(|h+y+(n-j-1)n^{-\frac{1}{2}}\hat{\mu}|^3|x|^3 + |x|^6) \\ + t^2(|h+y+(n-j-1)n^{-\frac{1}{2}}\hat{\mu}||x|^3 + |x|^4)]|g(t,0,(n-j-1)n^{-1}\hat{V})| . \end{aligned} \tag{47}$$

Applying Lemma 7 we have for $s \ge 3$

$$\begin{aligned} \int |x|^s \tilde{P}_{(n)}(dx) &= n^{-s/2}\int |x|^s \tilde{P}(dx) \\ &\le c(s)\beta n^{-3/2}. \end{aligned} \tag{48}$$

Denoting Y_1 a normal $(0,\hat{V})$ random variable and using (8) we also have

$$\begin{aligned} \int |x|^s \tilde{G}_{(n)}(dx) &= n^{-s/2}E|Y_1+\hat{\mu}|^s \\ &\le c(s)n^{-s/2}(E|Y_1|^s + |\hat{\mu}|^s) \\ &\le c(s)n^{-s/2}((E|Y_1|^2)^{s/2} + E|\hat{X}_1|^s) \\ &= c(s)n^{-s/2}((E|\hat{X}_1 - \hat{\mu}|^2)^{s/2} + E|\hat{X}_1|^s) \\ &\le c(s)n^{-s/2}(E|\hat{X}_1 - \hat{\mu}|^s + E|\hat{X}_1|^s) \\ &\le c(s)n^{-s/2}E|\hat{X}_1|^s \\ &\le c(s)\beta n^{-3/2} . \end{aligned} \tag{49}$$

Now

$$\int |h+y+(n-j-1)n^{-\frac{1}{2}}\hat{\mu}|^3 \tilde{P}_n^{(j)}(dy) = E|h+(n-1)n^{-\frac{1}{2}}\hat{\mu}+n^{-\frac{1}{2}}\sum_1^j (\tilde{X}_i-\hat{\mu})|^3$$

$$\leq c(s)(1+|h|^3+n^{-3/2}E|\sum_1^j(\tilde{X}_i-\hat{\mu})|^3) ,$$

since (cf. (40), (41))

$$\begin{aligned} |\hat{\mu}| &\leq E|\hat{X}_1|\chi_{A_1^c} \\ &\leq n^{-\frac{1}{2}}E|\hat{X}_1|^2 \\ &\leq n^{-\frac{1}{2}}\mathrm{tr}\, V . \end{aligned}$$

Furthermore

$$\begin{aligned} E|\sum_1^j (\tilde{X}_i-\hat{\mu})|^3 &\leq (E|\sum_1^j (\tilde{X}_i-\hat{\mu})|^4)^{3/4} \\ &= (E(\sum_1^j (\tilde{X}_i-\hat{\mu}), \sum_1^j (\tilde{X}_i-\hat{\mu}))^2)^{3/4} \\ &\leq c(jE|\tilde{X}_1-\hat{\mu}|^4+j^2(E|\tilde{X}_1-\hat{\mu}|^2)^2)^{3/4} , \end{aligned}$$

$$\begin{aligned} E|\tilde{X}_1-\hat{\mu}|^2 &= \mathrm{tr}\,\hat{V} \\ &\leq \mathrm{tr}\, V \end{aligned}$$

and by Lemma 7

$$\begin{aligned} E|\tilde{X}_1-\hat{\mu}|^4 &\leq cE|\tilde{X}_1|^4 \\ &\leq c\beta n^{\frac{1}{2}} , \end{aligned}$$

so that

$$E|\sum_1^j (\tilde{X}_i-\hat{\mu})|^3 \leq c(\beta+(\mathrm{tr}\ V)^2)^{3/4} n^{3/2} .$$

Consequently

$$\int |h+y+(n-j-1)n^{-\frac{1}{2}}\hat{\mu}|^3 \tilde{P}_n^{(j)}(dy) \leq c(S,\beta)(1+|h|)^3 . \tag{50}$$

From (45) - (50) we deduce

$$|d_n(t)| \leq c(S,\beta)(1+|h|)^3 t^2(1+|t|)n^{-3/2}\sum_{j=0}^{n-1}|g(t,0,(n-j-1)n^{-1}\hat{V})|$$

and similarly to (19) we may write (see Lemma 6)

$$\begin{aligned} \sum_{j=0}^{n-1}|g(t,0,(n-j-1)n^{-1}\hat{V})| &= 1+\sum_{j=1}^{n-1}|g(jn^{-1}t,0,\hat{V})| \\ &\leq 1+n\int_0^{(n-1)n^{-1}}|g(ut,0,\hat{V})|du \\ &\leq 1+nt^{-1}\int_0^{\infty}|g(v,0,\hat{V})|dv \\ &\leq 1+c(S)nt^{-1} . \end{aligned}$$

Hence

$$|d_n(t)| \le c(S,\beta)(1+|h|)^3 t^2(1+|t|)(1+nt^{-1})n^{-3/2} . \tag{51}$$

On the other hand by Lemmas 1,6,8

$$|d_n(t)| \le c(S)(1+t^2)^{-1} + \exp\{-c'n\}. \tag{52}$$

By Corollary 1 and Lemma 6 the random variable $|h+\hat{Y}|^2$ has a density bounded by $c(\sigma_1,\sigma_2,\sigma_3)$. Thus to estimate D_n we can apply Esseen's inequality which together with (51), (52) for any $Q,T,1<Q<T$ gives

$$\begin{aligned} D_n &\le c\left(\int_0^T |d_n(t)|t^{-1}dt + c(S)T^{-1}\right) \\ &\le c(S,\beta)(1+|h|)^3 n^{-3/2}\int_0^Q t(1+t)(1+nt^{-1})dt \\ &+ c(S)\left(\int_Q^\infty t^{-3}dt + T^{-1}\right) + cT\exp\{-c'n\} \\ &\le c(S,\beta)[(1+|h|)^3(Q^3 n^{-3/2} + Q^2 n^{-\frac{1}{2}}) \\ &+ Q^{-2} + T^{-1} + T\exp\{-c'n\}] . \end{aligned} \tag{53}$$

Assuming that

$$(1+|h|)^{3/2} n^{-\frac{1}{4}} \le 1 \tag{54}$$

we may put in (53) $Q = (1+|h|)^{-3/4} n^{1/8}$, $T = n^{\frac{1}{2}}$ and thus obtain

$$D_n \le c(S,\beta)(1+|h|)^{3/2} n^{-\frac{1}{4}} .$$

This proves the Theorem in the case (54). If (54) is not satisfied the Theorem is obviously true. □

Comments on Chapter II

The problem of estimation of the accuracy of normal approximation on balls with a fixed centre provided by the central limit theorem in Hilbert space was first considered in a special case by N.P. Kandelaki [16] (see also [44]) then by V.V. Sazonov [33] (in [33] the logarithmic speed of Kandelaki was improved to $n^{-1/6+\varepsilon}$, $\varepsilon>0$). Later a considerable progress was made by J. Kuelbs and T. Kurtz [18], who obtained the speed $n^{-1/6+\varepsilon}$, $\varepsilon>0$, under general moment conditions. The results of [18] were generalized and made more precise by V. Paulauskas [28], V.M. Zolotarev [46], V.V. Ulyanov [42,43] and others. The proof of Theorem 1,§1 given in the lectures uses a variant of the method of compositions and is based on the ideas of J. Kuelbs and T. Kurtz with modifications due to V. Paulauskas and V.V. Ulyanov. In the proof we avoided Fréchet derivatives (like J. Kuelbs and T. Kurtz in [18]) to make the presentation more elementary. Theorem 1,§2 belongs to V.V. Yurinskii [45]. Under additional assumptions (e.g. the independence of the components of X_1 in some basis) bounds with better speeds are also known (see e.g. [24,12]). In 47 F. Götze obtained the speed $n^{-\frac{1}{2}}$ assuming only $E|X_1|^6 < \infty$.

References

1. Bahr, B. von, On the central limit theorem in R^k, Ark. Mat. 7 (1967), 61-69.

2. ——— , Multi-dimensional integral limit theorems, Ark. Mat. 7(1967), 71-88.

3. Bergström, H., On the central limit theorem, Skand. Aktuarietidskr, 27(1944), 139-153.

4. ——— , On the central limit theorem in the space R_k, $k > 1$, Aktuarietidskr , 28(1945), 106-127.

5. ——— , On the central limit theorem in the case of not equally distributed random variables, Skand. Aktuarietidskr, 33(1949), 37-62.

6. Berry, A.C., The accuracy of the Gaussian approximation to the sum of independent variates, Trans. Amer. Math. Soc. 49(1941), 122-136.

7. Bhattacharya, R.N., On errors of normal approximation, Ann. Probab. 3(1975), 815-828.

8. Bhattacharya, R.N and Ranga Rao, R., Normal approximation and asymptotic expansions, Wiley, New York (1976).

9. Bhattacharya, R.N., Refinements of the multidimensional central limit theorem and applications, Ann. Probab. 5(1977), 1-27.

10. Bikelis, A., On the central limit theorem in R^k, Part I, II, Litovsk. Mat. Sb. 11(1971), 27-58; 12(1972), 53-84.

11. Blöndal, P.H., Explizite Abshätzung des Fehlers in Mehrdimensionalen Zentralen Grenzwertsatz, Dissert. Köln (1973).

12. Borovskih Yu V. and Račkauskas A., Asymptotic of distributions in Banach spaces, Litovsk. Math. Sb. 19(1979), 39-54.

13. Egglestone, H.G., Convexity, Cambridge University Press, Cambridge (1958).

14. Esseen, C.-G., On the Liapounoff limit of error in the theory of probability, Arkiv. Mat. Astr. Fysik, 28A (1942), 1-19.

15. ——— Fourier analysis of distribution functions. A mathematical study of the Laplace - Gaussian law, Acta Math. 77(1945), 1-125.

16. Kandelaki, N.P., On a limit theorem in Hilbert space, Trans. Comput. Centre Acad. Sci. Georgian SSR, 1(1965), 46-55.

17. Katz, M., Note on the Berry-Esseen theorem, Ann. Math. Stat. 34(1963), 1107-1108.

18. Kuelbs, J. and Kurtz T., Berry-Esseen estimates in Hilbert space and an application to the law of the iterated logarithm, Ann. Probab., 2(1974), 387-407.

19. Levy, P., Calcul de probabilités , Gautier - Villars, Paris (1925).

20. Liapounov, A.M., Sur une proposition de la théorie des probabilités Bull. Acad. Sc. St. Pétersbourg (5), 13(1900), 359-386.

21. ———, Nouvelle forme du théorème sur la limite de theorie des probabilités, Mem. Acad. Sci. St. Pétersbourg (8), 12(1901), 1-24.

22. Lindeberg,Y.W., Eine neue Herleitung des Exponentialgesetzes in der Wahrscheinlich keiterechnung, Math. Z., 15(1922), 211-225.

23. Nagaev,S.V., Some limit theorems for large deviations, Teor. Verojatnost. i Primenen, 10 (1965), 231-254.

24. Nagaev,S.V. and Čebotarev V.I., On estimates of the speed of convergence in the central limit theorem for random vectors with values in ℓ_2, Mathematical analysis and related topics, Nauka, Novosibirsk (1978), pp. 153-182.

25. Paulauskas,V., An improvement of Liapounov theorem, Litovsk. Mat. Sb., 9(1969), 323-328.

26. ———, On the rate of convergence in the multidimensional limit theorem with a stable limiting law, Litovsk. Mat. Sb., 15(1975), 207-228.

27. ———, An estimate of the remainder term in the multidimensional central limit theorem, Litovsk. Mat. Sb., 15(1975), 163-176.

28. ———, On the rate of convergence in the central limit theorem in some Banach spaces, Teor. Verojatnost. i Primenen., 21(1976), 775-791.

29. Rao, Ranga R., On the central limit theorem in R_k, Bull. Amer. Math. Soc., 67(1961), 359-361.

30. Rotar,V.I., A nonuniform estimate of the speed of convergence in the multidimensional central limit theorem, Teor. Verojatnost. i Primenen., 15(1970), 647-665.

31. ———, Non-classical estimates of the rate of convergence in the multi-dimensional central limit theorem, Part I, II, Teor. Verojatnost. i Primenen 22(1977), 774-790; 23(1978), 55-66.

32. Sazonov, V.V., On the multi-dimensional central limit theorem, Sankhyā Ser. A, 30 (1968), 181-204.

33. ———, An improvement of a convergence-rate estimate, Teor. Verojatnost. i Primenen., 14(1969), 667-678.

34. ———, On a bound for the rate of convergence in the multidimensional central limit theorem, Proceedings of the Sixth Berkeley symposium on Mathematical Statistics and Probability, Vol. II, University of California Press (1972), pp. 563-581.

35. ———, On the multidimensional central limit theorem with a weakened conditions on moments, Proceedings of the second Japan-USSR symposium on probability theory (Kyoto, 1972), pp. 384-396. Lecture Notes in Math. Vol. 330, Springer, Berlin (1973).

36. ———, A new general estimate of the rate of convergence in the central limit theorem in R^k, Proc. Nat. Acad. Sci. USA, 71(1974), 118-121.

37. Sazonov, V.V. and Ulyanov, V.V. On the speed of convergence in the central limit theorem, Adv. Appl. Prob. 11(1979), 269-270.

38. ——— , On the accuracy of normal approximation, submitted to Multivar. Anal.

39. Senatov, V.V., Several uniform estimates of the rate of convergence in the multidimensional central limit theorem, Teor. Verojatnost. i Primenen., 25(1980), 557-770.

40. Sweeting, T.J., Speed of convergence for the multidimensional central limit theorem, Ann. Probab. 5(1977), 28-41.

41. Ulyanov, V.V., A non-uniform estimate of the rate of convergence in the central limit theorem in R^k, Teor. Verojatnost. i Primenen., 21(1976), 280-292.

42. ———, Some improvements of convergence rate estimates in the central limit theorem, Teor. Verojatnost. i Primenen., 23(1978), 684-688.

43. ———, On the estimation of the rate of convergence in the central limit theorem in a real separable Hilbert space, Mat. Zametki, 28(1980), 465-473.

44. Vakhania, N.N. and Kandelaki, N.P., On the estimation of the speed of convergence in the central limit theorem in Hilbert space, Trans. Comput. Centre Acad. Sci. Georgian SSR., 10:1(1969), 150-160.

45. Yurinskii, V.V., On the error of the Gaussian approximation to the probability of the Gaussian approximation to the probability of hitting a ball, preprint.

46. Zolotorev, V.M., Ideal metrics in the problem of approximating the distributions of sums of independent random variables, Teor. Verojatnost. i Primenen., 22(1977), 449-465.

47. Götze, F., Asymptotic Expansions for Bivariate von Mises Functional, Z. Wahrscheinlichkeitstheorie verw. Gebiete 50, (1979), 333-355.

Index